Yamato
VS
Iowa
Written & Illustrated by Paul Forest

Copyright © 2021 Paul Forest
All rights reserved.
ISBN: 9798744613235

Introduction

On 4 November 1937 the super battleship Yamato was secretly laid down at Kure Naval Arsenal. Less than three years later, on 27 June 1940, invoking the "escalator clause", the Americans laid down the first unit of their greatest battleship class, USS Iowa, at the New York Naval Yard. Lacking accurate intel, both navies were convinced that their newest battleship was second to none, and thus capable of overwhelming any foe in a gunnery duel. In the possession of the actual technical characteristics of the two ships, relying on primary sources and empirical data, we now take on, once again, one of the most hotly debated questions among naval analysts and enthusiasts ever since: Who was right?

Contents

Introduction	5
Abbreviations	9
Chapter 1 – Firepower	11
Iowa's Main Battery	11
Yamato's Main Battery	13
Comparison of Main Batteries	15
Secondary Armament	26
Chapter 2 – Fire-Control	27
Iowa's Main Battery Fire-Control	27
Yamato's Main Battery Fire-Control	29
Comparison of Main Battery Fire-Control Systems	31
Chapter 3 – Protection	43
Iowa's Protection Scheme	43
Yamato's Protection Scheme	46
Comparison of Protection Schemes	48
Conclusion	73
Appendix A – General Characteristics (WWII)	83
Appendix B – Calculating Penetration Values	85
Appendix C – Japanese Armour Compared to Foreign Types	91
Appendix D – U.S. Armour	97
Appendix E – Flat Penetrators	99
Appendix F – Penetrator Hardness	101
Appendix G – Radars	103
Appendix H – Iowa's Mark 37 Fire-Control System	117
Appendix I – Yamato's Armour Decks	121
Appendix J – U.S. Battleship's Combat Information Center (CIC)	123
Appendix K – Yamato's Main Turrets	125
Appendix L – Projectiles	129
Appendix M – Longitudinal Sections	135
Definitions	137
List of Tables	141
List of Figures	143
List of Pictures	145
Bibliography	147

Abbreviations

A – Class A (U.S. cemented armour)
ADM – Admiralty Minutes
AP – Armour Piercing (projectile)
ATC – Armour Technical Committee
B – Class B (U.S. non-cemented armour)
Bhd. – Bulkhead
BHN – Brinell Hardness Number
BL – Ballistic Limit (velocity)
Bu. Ord. – Bureau of Ordnance
C – Cemented (armour) or Coincidence (rangefinder)
cal. – caliber(s)
CIC – Combat Information Center
CNC – Copper Non-Cemented (Japanese light non-cemented armour)
crh – caliber radius head
CT – Conning Tower
deg. – degree(s)
DP – Dual Purpose (gun)
DS – Ducol Steel
FH – Face-Hardened
Fwd. – Forward
GFCS – Gun Fire-Control System
HTS – High Tensile Steel
IFF – Identification Friend or Foe
IV – Initial Velocity
LOS – Line of Sight
MNC – Molybdenum Non-Cemented (Japanese non-cemented armour)
Mod. – Model
NC – Non-Cemented (armour)
NPG – Naval Proving Ground
NVNC – New Vickers Non-Cemented (Japanese non-cemented armour)
O – Ordnance (USNTMJ)
OP – Ordnance Pamphlet
OT – Ordnance Targets (USNTMJ)
PPI – Plan Position Indicator
PRF – Pulse Repetition Frequency (radars)
Rc – Rockwell C (hardness)
RE – Relative Effectiveness (explosives)
RF – Rangefinder
RPC – Remote Power Control
S – Stereoscopic (rangefinder)
SH – Shore Hardness
Sk. – Sketch

STS – Special Treated Steel (U.S. thin non-cemented armour)
T/D – Thickness/Diameter (armour/penetrator)
USNTMJ – U.S. Naval Technical Mission to Japan
VC – Vickers Cemented (Japanese cemented armour)
VH – Vickers Hardened (Japanese face-hardened armour)

Chapter 1 – Firepower

Iowa's Main Battery

In the 1930s the Americans introduced new types of naval armour-piercing projectiles with an extremely high density factor (~0.66 lbs/in^3). In order to achieve this, projectile bodies were lengthened, and the weight of their bursting charge was diminished to only 1.5 % of that of the projectile. The filling material of the new shells was Dunnite, an explosive substance with a relative effectiveness factor of approximately 0.95. The major calibre projectiles were "sheath hardened" and had a maximum hardness of about 555 BHN. The armour-piercing caps had the same max. hardness. Like the shells themselves, these were robust and relatively blunt too, especially the later mods. Although performance at high obliquity was very good, post-war tests revealed that the projectiles would have been more effectual had their maximum hardness been greater, as was the case with smaller (6"/152 mm and 8"/203 mm) projectiles. All AP type projectiles developed for modern 6", 8", 12" (152 mm, 203 mm, 305 mm), and 16" (406 mm) guns had a uniform length of 4.5 cal. The largest ones had 4.5/9 crh ballistic caps.

Firing "super heavy" type shells, modern American naval guns had low muzzle velocity and inferior ballistic characteristics. However, the new shells were exceedingly resistant to retardation, due to their very high sectional density. The immense terminal energy of the new projectiles resulted in outstanding piercing capacity, especially in the case of horizontal armour, against which their plunging trajectories made them particularly effective.

Iowa's guns propelled the same "super heavy" type projectiles that were originally developed for the preceding 16" (406 mm)/45 Mark 6 guns mounted on the North Carolina and South Dakota battleship classes. However, the new gun was 50 cal. long, and its average pressure was raised by 7 %. As a result of these modifications, muzzle velocity increased from 2,300 fps (701 mps) to 2,500 fps (762 mps). Consequently, the new gun had superior ballistic characteristics and vertical armour-piercing capacity, but their barrel life and horizontal armour-piercing capacity were diminished. This made the new guns more balanced than their predecessor, which had extraordinary horizontal armour-piercing capacity and durability, but poor ballistic characteristics.

There was little difference between the two guns in terms of performance beyond what has already been stated. The loading cycles of the Mark 6 and the Mark 7 guns were both established at 30 seconds, although this flattering value was not achieved in action. The highest documented sustainable rate of fire attained by modern U.S. 16" (406 mm) guns in action was circa 1.5-1.6 rounds/gun/minute – this was achieved by USS Washington during her engagement with the Japanese battlecruiser Kirishima, at point-blank range, with the guns at very low elevation. According to Bulletins of Ordnance information published in 1944, dispersion was 1.0 % of range for a three-gun salvo and 1.9 % for a nine-gun broadside for both the Mark 6 and the Mark 7 guns.

Picture 1.1

Exquisite view of USS New Jersey, 11 September 1968.

Yamato's Main Battery

Like the Americans, the Japanese also developed new types of AP projectiles for the Pacific War. The design of the new projectiles was greatly influenced by the firing trials carried out against the cancelled battleship, Tosa — a full-scale ordnance test revealing the vulnerability of armoured ships against underwater hits. Impressed by the results, the Japanese designed projectiles with enhanced underwater performance. This was obtained by separating the armour pricing cap into two elements, namely a small upper section, or cap head, and a more robust lower section, or cap proper. The contact area of the two was completely flat and the cap head was attached to the windshield of the shell. Upon contact with water or armour at high obliquity, the cap head broke off, along with the ballistic cap. The resulting flat contact surface of the cap stabilized the projectile's trajectory underwater, increasing its danger space by approximately 100-200 cal.

Much less acclaimed is the fact that, according to NPG reports No. 7-43 and No. 4-44, the completely flat nose of the cap, which made contact with armour at high obliquity, made the projectile exceptionally efficacious against low T/D ratio targets — i.e. armour decks or tapering armour carried below the waterline — for two reasons. First, flat-nosed projectiles experience a righting torque bringing them towards the plate normal during the initial phase of perforation — not away from it, as is the case with pointy projectiles. Second, blunt rigid penetrators achieve perforation by plugging, a defeat mechanism that requires less energy than piercing, the defeat mechanism associable with sharp rigid penetrators.

Equally beneficial was the excessive surface hardness characterizing the new Japanese projectiles according to NPG reports No. 4-44 and No. 3-47, as well as the document titled *Effect of Projectile Nose Shape on Ballistic Limit Velocity, Residual Velocity, and Ricochet Obliquity*. Based on USNTMJ OT O-19 *(Japanese Projectiles General Type)*, hardness distribution was "detrimental" and the maximum hardness of the new Japanese AP projectile bodies and armour-piercing caps ranged from 80 SH (634 BHN) to 83 SH (670 BHN).

As might be expected, the official armour-piercing capacity values of Yamato, surpassing the predictions of Bu. Ord. Sk. 78841 by a wide margin, are extraordinary — especially against horizontal armour at relatively short range — even for a gun of this caliber. A number of naval analysts trace this back to the allegedly inferior ballistic performance of Japanese armour as compared to foreign types; but, in light of empirical data, it seems more likely that the reasons for this were those stated above. The mean performance of Japanese armour plates tested post-war cannot justify the assumption that this nation produced armour plates of markedly inferior quality in general, whereas the superiority of flat/hard penetrators (under certain conditions) is backed by consistent empirical data.

The new projectiles had an average density factor (0.55 lbs/in^3). As filling material, the Japanese adopted TNA, a rather potent substance with an R.E. of 1.05. The filler of the 18.07" (459 mm) projectile was the largest among naval AP shells at the time, although percentage-wise (2.3 %) it was smaller than those of British and French AP shells. The ballistic cap was not ogival, but a cone with a half nose angle of 11.75 deg. This was later abandoned in favor of a longer, more tapering cap with a half nose angle of only 10.5 deg., increasing the length of the projectile from 4.31 cal. to 4.47 cal. The new projectiles also had a revised driving band. Both types were boat-tailed.

The guns had moderate average pressure. As a result, muzzle velocity was not particularly high, though it was higher than those of contemporary U.S. and British guns. However, owing to the high maximum elevation of the guns, and the superior underwater performance of projectiles, their maximum range and danger space were outstanding.

The reloading time of the guns proved to be very favorable too. Based on USNTMJ OT O-45 (N) *(Japanese 18" Guns and Mounts)*, a minimum reload time of 28 seconds (2.14 rounds/gun/min.) could be achieved at low elevation, albeit it was concluded that sustainable reload time at optimum battle range would have been circa 35 seconds (1.71 rounds/gun/min.). According to the same document, the guns and mounts proved to be very reliable. On average, dispersion was 1.2 % of range with the guns at maximum elevation.

Picture 1.2

The gargantuan forward turrets of Musashi, photographed in 1942.

Comparison of Main Batteries

The Americans attained remarkable terminal energies by maximizing the density factor of their new shells, but this was countered on the Japanese side by larger and harder projectiles, as well as a more favorable nose shape against deck armour. Not taking mechanical properties and nose shape into consideration, Bu. Ord. Sk. 78841 predicts only mildly superior penetration capacity for Yamato's guns, but the official penetration values of the Japanese gun are at least as much above those of its U.S. analogue as might be expected from a gun of this much greater caliber.

The rate of fire, dispersion, and ballistic characteristics of the two guns were comparable, with Yamato's having a slight edge in all three categories. However, the danger space of the Japanese projectiles was much larger by virtue of the higher emphasis placed on their underwater performance.

The Americans sacrificed explosive content so as to amplify projectile weight, and this inevitable shortcoming was not diminished by the adoption of a comparatively weak explosive material. Yamato's projectile, carrying a larger quantity of a more potent explosive, was more destructive.

Table 1.1

Armament Comparison					
Designation	Iowa			Yamato	
Gun)					
Muzzle Energy (MJ)	356			450	
Muzzle Velocity (mps/fps)	762	2,500		785	2,575
Maximum Range (m/yds)	38,720	42,345		42,030	45,965
Min. Reload Time (sec)	30			28	
Dispersion (% of range)	1.0-1.9*			1.1-1.3**	
AP Projectile)					
Weight (kg/lbs)	1,225	2,700		1,460	3,219
Bursting Charge (kg/lbs/%)	18.55	40.9	1.51	33.85	74.63	2.32
Length (mm/in/cal.)	1,829	72	4.5	1,980-2,054***	77.95-81.0***	4.31-4.47***
Density Factor (lbs/in^3)	0.66			0.55	
Sectional Density (lbs/in^2)	10.58			9.86	
Caliber Radius Head	4.5/9			6/∞	
Hardness Distribution	'Sheath Hardened'			'Detrimental'	
Maximum Hardness (BHN)	555			670	
Notes)					
*=1.0 %=3-gun salvo, 1.9 %=9-gun salvo, based on Bulletins of Ordnance, 1944. **=500-600 yds (457-549 m) at max. elevation (range=45,965 yds/42,030 m), based on USNTMJ OT O-45 (N) (*Japanese 18" Guns and Mounts*). Max. rate of fire is also based on this document. ***=Type 91-Type 1.					

Table 1.2

Broadside Firepower Comparison				
Designation	Iowa		Yamato	
Energy (MJ)	3,200		4,050	
Weight (kg/lbs)	11,022	24,300	13,140	28,969
Bursting Charge(kg/lbs)	167	368	305	672
Energy/min. (MJ)	6,400		8,679	
Weight/min. (kg/lbs)	22,044	48,600	28,157	62,076
Bursting Charge/min. (kg/lbs)	334	736	654	1,442

Table 1.3

Ballistic Characteristics									
Range		Angle of Descent (deg.)		Danger Space for a 20 ft (6.1 m) target – yds (m)		Striking Velocity – fps (mps)		Kinetic Energy (MJ)	
kyd	km	Iowa	Yamato	Iowa	Yamato*	Iowa	Yamato	Iowa	Yamato
5.5	5	2.8	3.3	136.3 (124.7)	115.6 (105.8)	2,258 (688)	2,264 (690)	290	348
10.9	10	6.4	7.2	59.4 (54.4)	52.8 (48.3)	2,039 (621)	2,034 (620)	237	281
16.4	15	11.1	11.5	34.0 (31.1)	32.8 (30.0)	1,846 (563)	1,844 (562)	194	231
21.9	20	17.2	16.5	21.5 (19.7)	22.5 (20.6)	1,694 (516)	1,709 (521)	163	198
27.3	25	24.4	23.0	14.7 (13.4)	15.7 (14.4)	1,596 (486)	1,608 (490)	145	175
32.8	30	32.6	31.4	10.4 (9.5)	10.9 (10.0)	1,554 (474)	1,558 (475)	137	165

*= + 50-100 yds (46-92 m) – Diving
Note= Yamato probably fires Type 91 projectile at a velocity of 2,559 fps (780 mps). The Type 1 projectile most likely had better ballistic characteristics due to its longer and more streamlined ballistic cap.

Table 1.4

Range		Time of Flight (sec)	
yds	m	Iowa	Yamato
18,406	16,830	26.7	26.1
30,534	27,920	51.6	49.2
39,184	35,830	76.7	70.3

Table 1.5

Armour Penetration Capacity Based on Bu. Ord. Sk. 78841* – in (mm)							
Range		Vertical Armour at 90 deg.		Vertical Armour at 60 deg.		Horizontal Armour	
kyd	km	Iowa	Yamato	Iowa	Yamato	Iowa	Yamato
0	0	32.64 (829)	34.07 (865)	---	---	---	---
5.5	5	29.08 (739)	29.53 (750)	---	---	---	---
10.9	10	25.60 (650)	25.69 (653)	---	---	---	---
16.4	15	22.17 (563)	22.25 (565)	17.11 (435)	17.26 (438)	3.09 (78)	3.22 (82)
21.9	20	18.99 (482)	19.43 (494)	14.86 (377)	15.25 (387)	4.37 (111)	4.29 (109)
27.3	25	16.20 (411)	16.73 (425)	12.89 (327)	13.33 (339)	5.84 (148)	5.62 (143)
32.8	30	13.79 (350)	14.22 (361)	11.18 (284)	11.53 (293)	7.65 (194)	7.46 (189)

*=The official penetration curves of Iowa are based on this formula.

Table 1.6

Armour Penetration Capacity Based on and Bu. Ord. Sk. 78841* (Iowa) and Official Values (Yamato)** – in (mm)					
Range		Vertical Armour at 90 deg.		Horizontal Armour	
kyd	km	Iowa	Yamato	Iowa	Yamato
21.9	20	18.99 (482)	22.28 (566)	4.37 (111)	6.57 (167)
32.8	30	13.79 (350)	16.38 (416)	7.65 (194)	9.06 (230)

*=The official penetration curves of Iowa are based on this formula.
**=These values are in substantial agreement with USNTMJ OT O-19 *(Japanese Projectiles General Type)*. According to this document, one of the test specifications of the 18.07" (459 mm) projectiles was to defeat 16.14" (410 mm) armour plates at 30 deg./1,723 fps (525 mps). However, it was claimed that test specification velocities for the 18.07" (459 mm) projectile were approximately 200 fps (61 mps) above ballistic limit velocity, i.e. ballistic limit velocity was approximately 1,523 fps (464 mps) for a 16.14" (410 mm) plate at 30 deg. Now, the angle of descent and striking velocity of the 18.07" (459 mm) projectile at 32.8 kyd (30 km) is 31.4 deg. and 1,558 fps (475 mps), respectively. The piercing capacity given above (16.38"/416 mm) is, therefore, in very good agreement with these values. It should also be noted that, according to the same document, 22.05" (560 mm) plates were attacked at 16.5 deg., which is exactly the same as the angle of descent of the projectiles at 21.9 kyd (20 km). Unfortunately, striking velocity is specified by the document only for the 16.14" (410 mm) plate.

The ship's designed period of immunity is also in good agreement with these values, with the inner limit of immunity established at 21.9 kyd (20 km) and her vertical armour (barbette and CT sides) having a thickness of 22.05" (560 mm). The outer limit of immunity was at 32.8 kyd (30 km). With the effective resistance of all the ship's protective decks combined closely approximating that of a 9.06" (230 mm) thick single plate, this horizontal piercing capacity at this range is also in good agreement with the designed immunity of the ship.

Figure 1.1

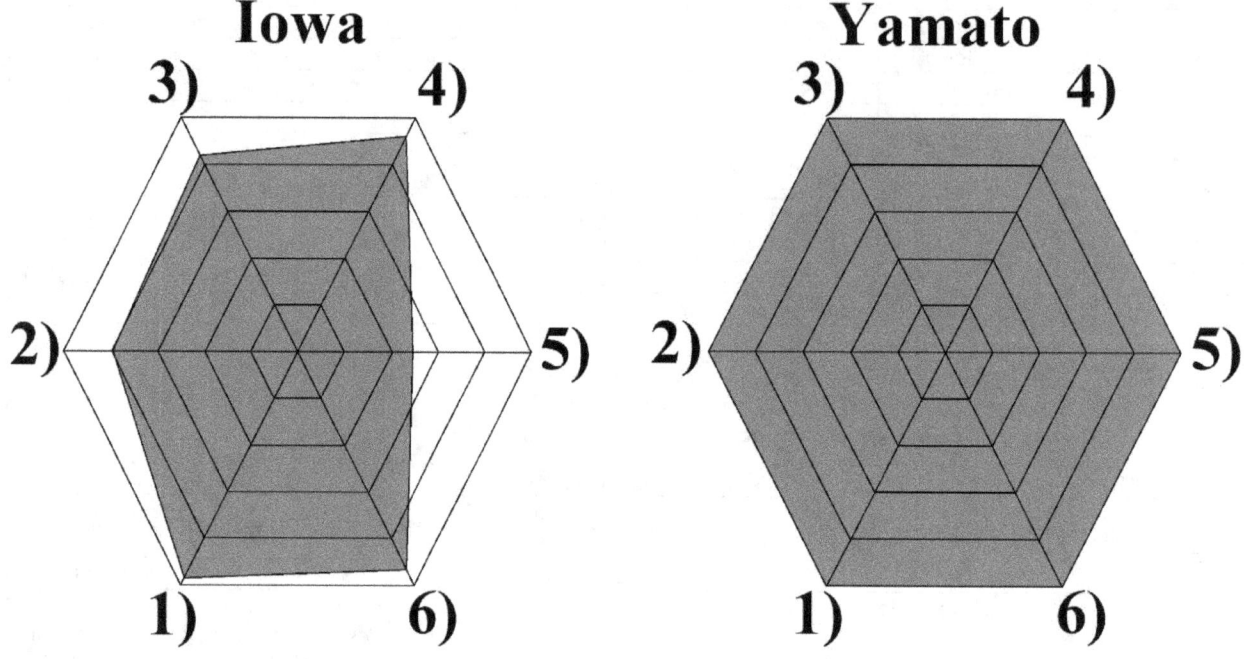

Figure 1.2

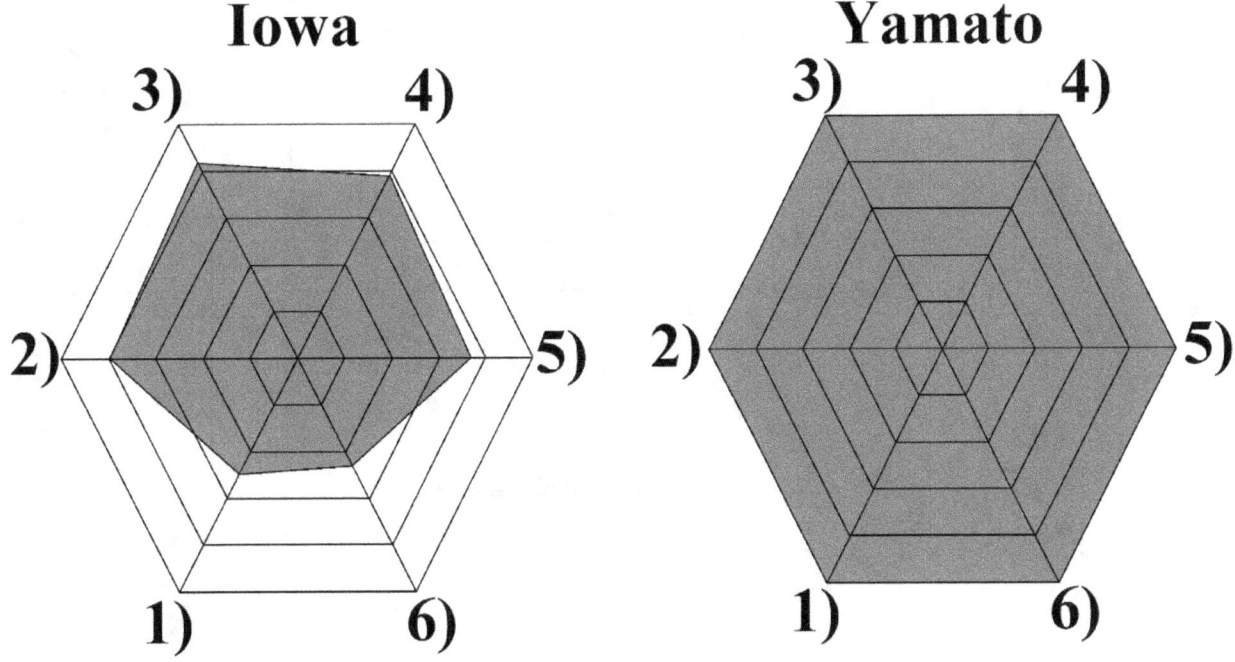

1) Bursting Charge   4) Weight/min.
2) Energy   5) Energy/min.
3) Weight   6) Bursting Charge/min.

Figure 1.3

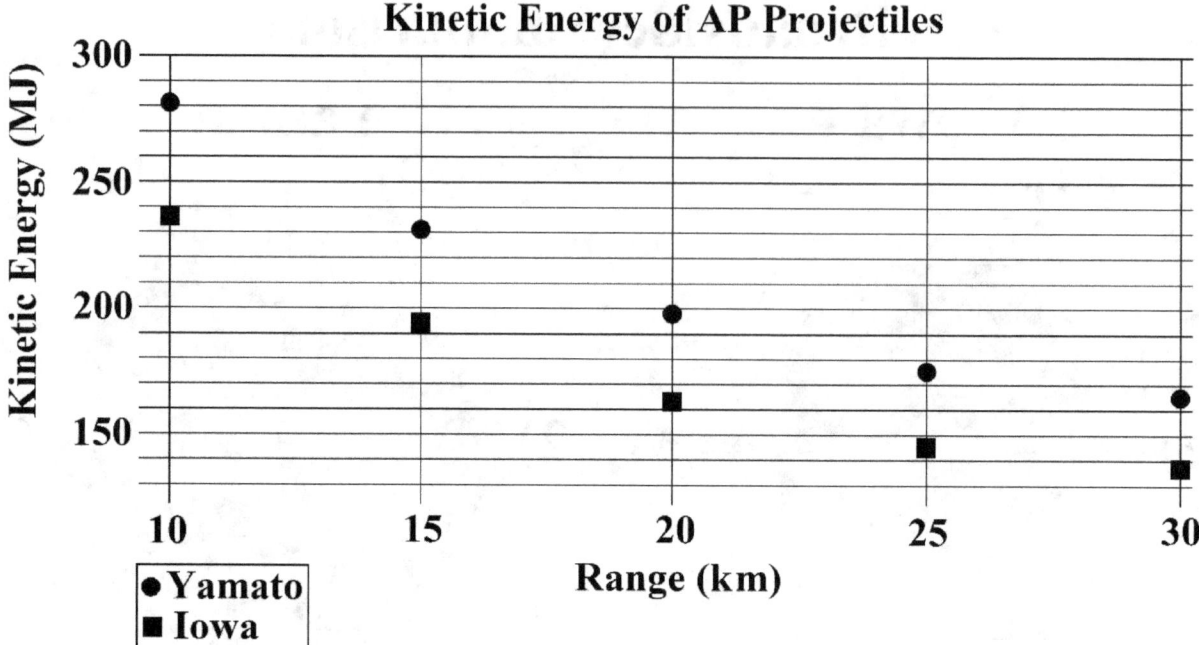

Figure 1.4

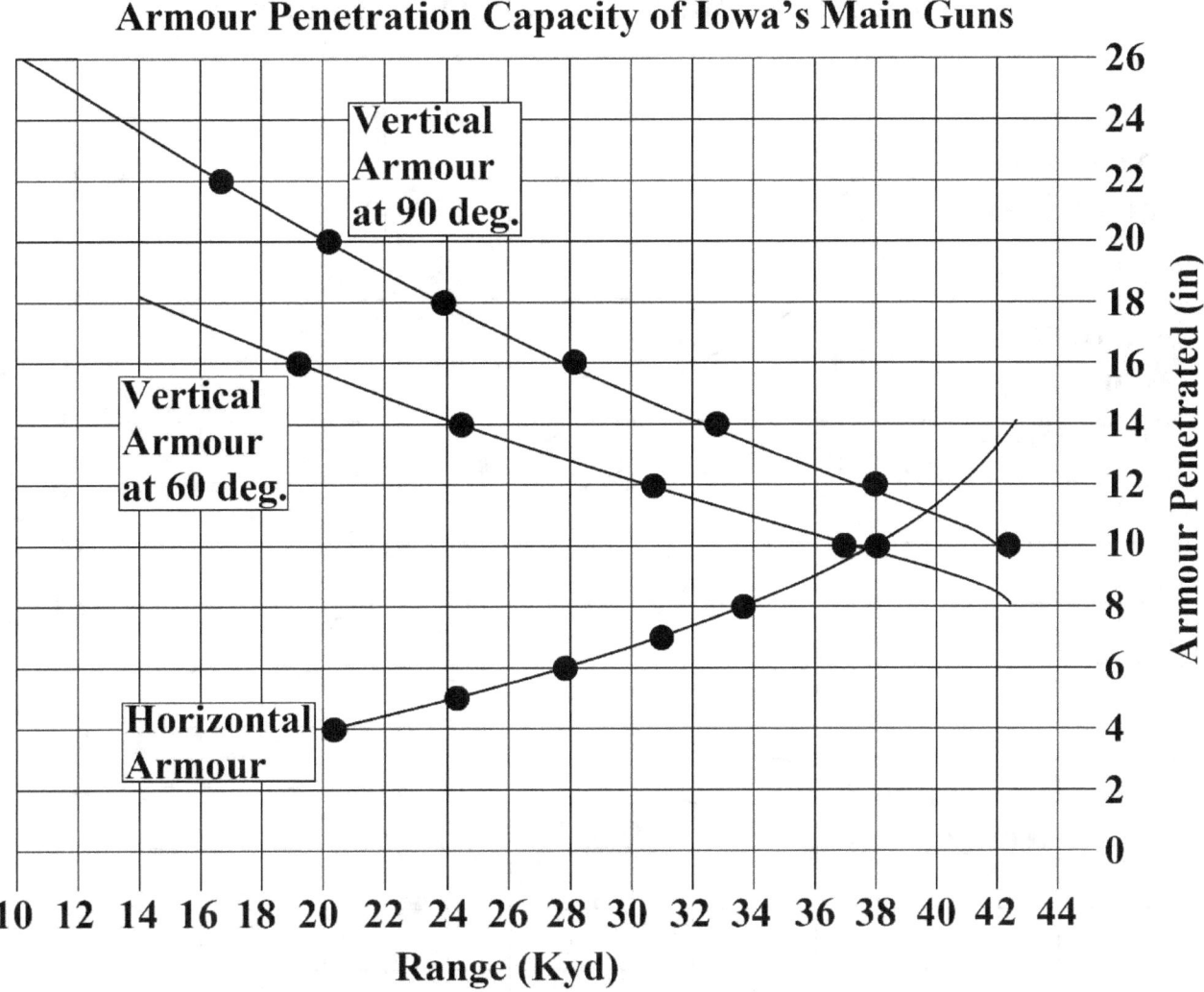

Figure 1.5

Armour Penetration Capacity of Yamato's Main Guns

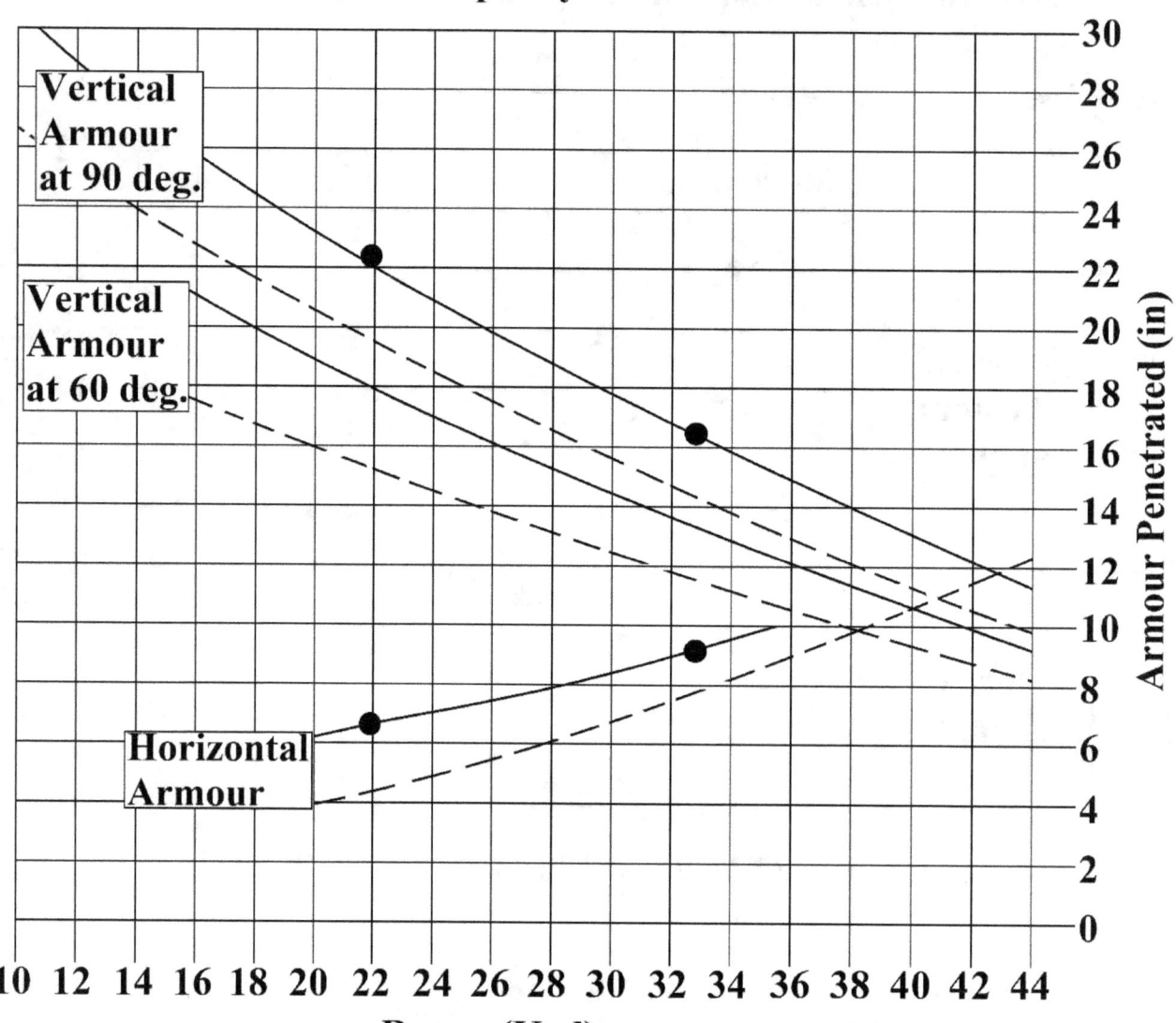

Description of Figure 1.5:

The official vertical armour penetration values of Yamato are probably based on the DeMarre formula:

$$t=[v\cdot\cos(\Theta)\cdot m^{1/2}/(K\cdot d^{3/4})]^{10/7}$$

Whence
t=thickness {mm},
v=velocity {mps},
m=mass {kg},
d=diameter {mm},
Θ=obliquity {deg.},
K=coefficient – vertical armour {2.28}.
K=coefficient – horizontal armour {2.25-0.036*(35-Θ)}.

Yamato's Penetration Capacity in inches (mm)				
Range	Vertical Armour at 90 deg.		Horizontal Armour	
kyd (km)	DeMarre (K=2.28)	Official	DeMarre (K=2.25-0.036*(35-Θ))	Official
0	42.05 (1068)	---	---	---
5.5 (5)	35.20 (894)	---	1.69 (43)	---
10.9 (10)	29.96 (761)	---	3.70 (94)	---
16.4 (15)	25.59 (650)	---	5.28 (134)	---
21.9 (20)	**22.24 (565)**	**22.28 (566)**	6.57 (167)	6.57 (167)
27.3 (25)	19.21 (488)	---	7.80 (198)	---
32.8 (30)	**16.50 (419)**	**16.38 (416)**	9.06 (230)	9.06 (230)

The K coefficient we gave for horizontal armour penetration, although in very good agreement with official data, gives plausible values only up to about 35 kyd (32 km). Beyond this range, it probably underestimates piercing capacity.

Note that the DeMarre formula indicates considerably greater piercing value as compared to Bu. Ord. Sk. 78841, especially against horizontal armour at relatively short ranges:

Yamato's Penetration Capacity in inches (mm)				
Range	Vertical Armour at 90 deg.		Horizontal Armour	
kyd (km)	DeMarre	Bu. Ord. Sk. 78841	DeMarre	Bu. Ord. Sk. 78841
0	42.05 (1068)	34.07 (865)	---	---
5.5 (5)	35.20 (894)	29.53 (750)	---	---
10.9 (10)	29.96 (761)	25.69 (653)	---	---
16.4 (15)	25.59 (650)	22.25 (565)	5.28 (134)	3.22 (82)
21.9 (20)	22.24 (565)	19.43 (494)	6.57 (167)	4.29 (109)
27.3 (25)	19.21 (488)	16.73 (425)	7.80 (198)	5.62 (143)
32.8 (30)	16.50 (419)	14.22 (361)	9.06 (230)	7.46 (189)

This is probably due to the fact that the coefficient in the U.S. formula is adjusted to softer (555 BHN) projectiles with conventional AP caps.

Figure 1.6

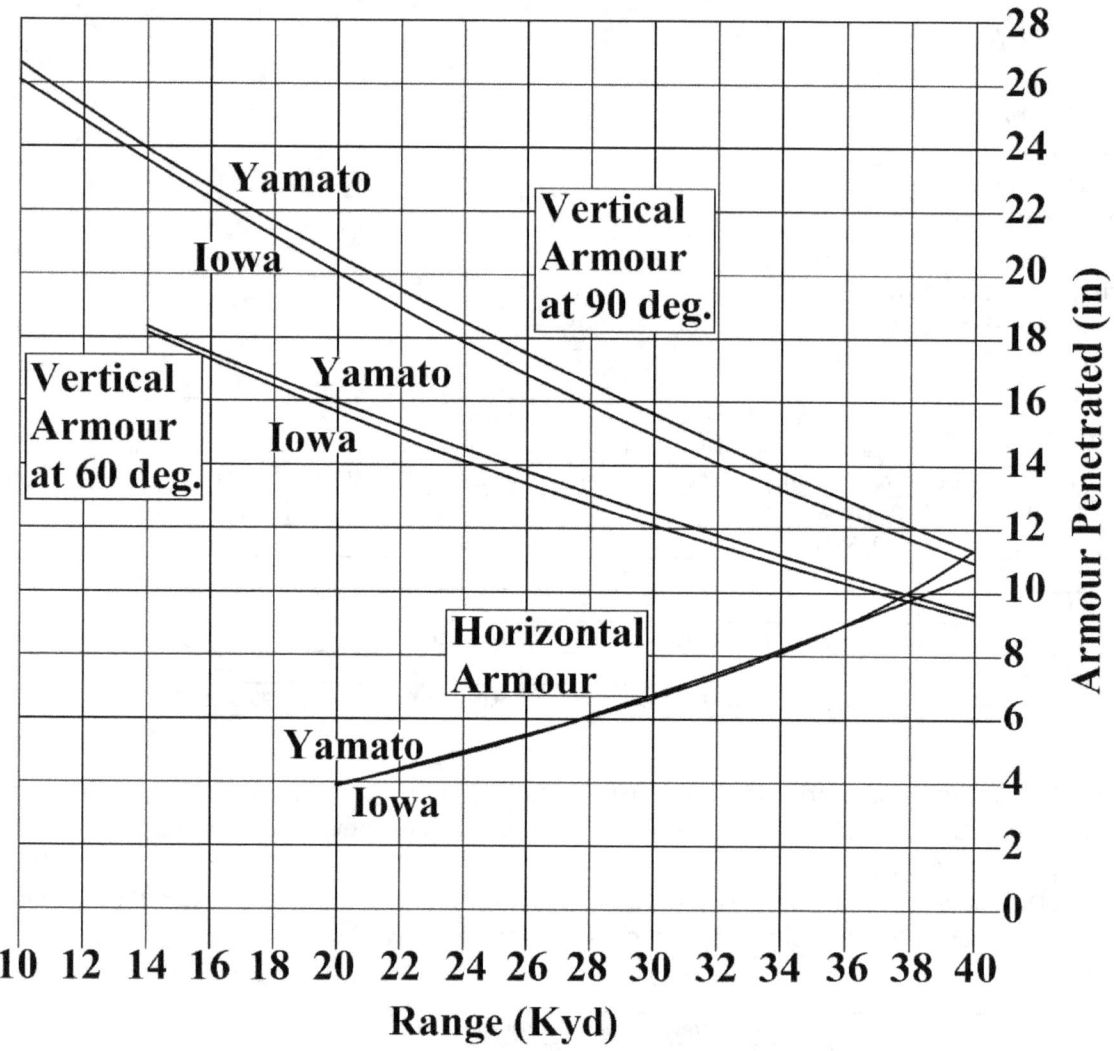

Figure 1.7

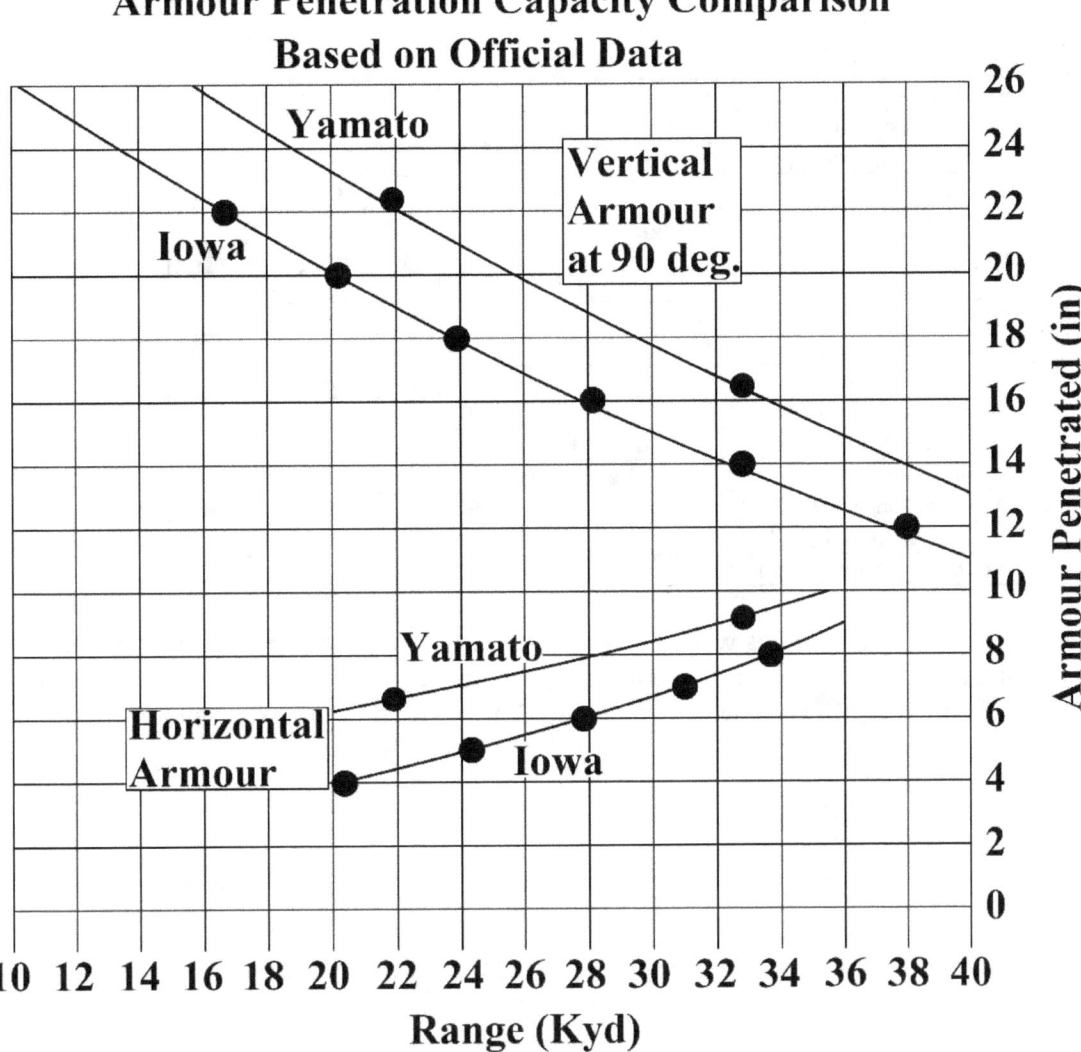

Secondary Armament

Table 1.7

Armament Comparison				
Designation	Iowa		Yamato	
Low-Angle Guns)				
No. of Barrels/Side			9 (6)*	
Muzzle Energy (MJ)			23.9	
Muzzle Velocity (mps/fps)			925	3,035
Projectile weight (lbs/kg)	---		55.87	123.2
Bursting Charge (lbs/kg)			3.1	6.8
Maximum Range (m/yds)			27,400	29,965
Rate of Fire (rounds/gun/min)			5-7	
DP Guns)				
No. of Barrels/Side	10		6 (12)*	
Muzzle Energy (MJ)	7.9		6.2	
Muzzle Velocity (mps/fps)	792	2,600	725	2,379
Projectile weight (lbs/kg)	25.0	55.2	23.45	51.7
Bursting Charge (lbs/kg)	1.2	2.6	1.9	4.2
Maximum Range (m/yds)	15,903	17,392	14,800	16,185
Rate of Fire (rounds/gun/min)	12-22		8-14	
*=As built (as modernized)				

Notwithstanding that the U.S. DP guns were more successful than the Japanese ones, it is clear that the secondary armament of Yamato was more efficient in harassing enemy surface targets due to her high-performance 6.1" (155 mm)/60 guns, which had no equivalents on the American side.

Chapter 2 – Fire-Control

Iowa's Main Battery Fire-Control

The Americans preferred stereoscopic rangefinders, albeit the Mark 53 rangefinder mounted on the No. 1 main gun turret was of coincidence type. This instrument was later suppressed, however, leaving only stereoscopic rangefinders for optical ranging.

All primary armament and DP gun directors of these ships were equipped with fire-control radars from their initial commission.

A total of three directors provided a constant flow of target information for the ship's main battery, namely two Mark 38 directors, one atop the aft and the other atop the forward, fire-control tower, plus one Mark 40 director, which occupied the upper level of the conning tower. Optical sensors of the Mark 38 directors included a 26.5 ft (8.08 m) Mark 48 stereoscopic rangefinder and two Mark 69, one Mark 56 and one Mark 29 telescopes. Each director was also fitted with a Mark 8 gunnery radar, which was succeeded by a more advanced set after World War II. The Mark 40 director incorporated two Mark 30 and one Mark 32 telescopes and one Mark 3 radar. The latter was replaced by the superior Mark 27 radar in the final year of World War II. The mark 40 director housed no rangefinder.

The foretop director was 116 ft (35.4 m) above the waterline, which corresponds to a visual/radar horizon of 23.2 kyd/26.8 kyd (21.2 km/24.5 km) and 31.7 kyd/36.6 kyd (29.0 km/33.5 km), assuming a target height of zero and 100 ft (30.5 m), respectively.

All main turrets incorporated four Mark 66 and two Mark 28 or 29 telescopes. Both the No. 2 and No. 3 gun turrets mounted a 46 ft (14.02 m) Mark 52 stereoscopic rangefinder. These were stabilized by Mark 3 rangefinder stabilizers. As depicted above, turret No. 1 was equipped with a 46 ft (14.02 m) Mark 53 coincidence rangefinder, but this instrument was later removed. Both functions of all three turrets could be remotely controlled, but a Mark 3 computer was also allocated for local fire-control.

The Iowa class battleships had two main battery plotting rooms, both inside the armoured citadel. Both the forward and the aft potting rooms contained a Mark 8 rangekeeper, i.e. analogue fire-control computers for synthetic plotting. A Mark 41 stable vertical was also incorporated in each plotting room. This device compensated for the inherent movements of the ship. A Mark 21 fire-control switchboard, which controlled the circuits of the primary armament fire-control system, was also provided. The ship's speed and course were measured by the ship's pitometer log and gyro compass, respectively.

As commissioned, Iowa carried a total of three Vought OS2U Kingfisher floatplanes for reconnaissance and splash spotting, but these were superseded by Curtiss SC Seahawks during the final year of the war.

Presupposing good visibility, spotting preference was as follows:

Order of Preference	Range	Deflection
1	Radar	Optical
2	Air	Radar
3	Optical	Air

The ship could use a total of five 36 in (91.4 cm) searchlights, while the DP guns could fire starshells, to illuminate targets at night.

Picture 2.1

Iowa's aft Mark 37 (5"/127 mm) and Mark 38 (16"/406 mm) directors, captured on 8 October 1983. The World War II period Mark 4 and Mark 8 fire-control radars had been replaced by the more advanced Mark 25 and Mark 13 sets, respectively. Notice the protruding hoods of the rangefinders.

Yamato's Main Battery Fire-Control

In the new battleship design, all three main gun turrets and both main battery directors (hoiban) mounted extraordinarily large triplex rangefinders, incorporating one stereoscopic- and two coincide-type optical systems each. The base length of all turret rangefinders and that of the foretop director was as much as 49.21 ft (15.0 m), whereas the aft director's rangefinder was 32.81 ft (10.0 m) long. The foretop rangefinder was 122.4 ft (37.3 m) above the waterline, giving a visual horizon of 23.8 kyd (21.8 km) and 32.1 kyd (29.4 km) for a dimensionless and a 100 ft (30.5 m) high target, respectively.

Each side of Yamato's forward fire-control tower was fitted with a Type 22 mod. 4 combined surface search/gunnery radar. Even though this instrument was quite rudimentary, its detection range and range accuracy were reasonably good. The antenna was about 107 ft (32.5 m) above the waterline, yielding a radar horizon of 25.7 kyd (23.5 km) and 35.8 kyd (32.7 km) for a dimensionless and a 100 ft (30.5 m) high target, respectively.

The fire-control problem was solved by the Type 98 fire-control system, which was specifically designed for the new battleships. Among other things, and the sensors introduced above, the Type 98 system incorporated target speed and bearing indicator panels (shokutekiban) below each director, and a below-deck range-averager (sokkyo heikinban), to which the range measurements were relayed before their mean value was transmitted to the fire-control computer (shagekiban). A gyro horizon compensated for roll and pitch. Dispersion was minimized by a firing time limiting device, which prevented shell interference. Speed was measured by the ship's pitometer log, aloft atmospheric conditions by a balloon.

Based on USNTMJ OT O-31 *(Japanese Surface and General Fire Control)*, the Type 98 system had the following limits:

Table 2.1

Limits of the Type 98 Fire-Control System	
Measured Range	54,681 yds (50,000 m)
Gun Range	45,166 yds (41,300 m)
Maximum Deflection	Right 130 mils, Left 160 mils
Deflection in Azimuth	500 mils
Own Ship Speed	35 knots
Enemy Ship Speed	40 knots
Wind Speed	131 fps (40 mps)

The ship initially carried a total of eight 59.1" (150 cm) searchlights, albeit their number was later reduced. Starshells were developed for both the 5" (127 mm) and the 6.1" (155 mm) guns.

The ship could accommodate up to seven floatplanes of the Mitsubishi F1M or Aichi E13A types for reconnaissance and shot spotting, but in practice only the former type was carried.

Picture 2.2

Excellent view of the fire-control tower of Musashi, captured in 1942. The rangefinder hoods of the main and secondary turrets, as well as that of the foretop director, can be clearly seen.

Comparison of Main Battery Fire-Control Systems

Little to no relevant empirical data is at our disposal that could help us reach a meaningful conclusion regarding the comparative performance of the fire-control systems of the two ship, but it is safe to say that the more powerful radar equipment of Iowa was better suited for fire-control, although this was, to some extent, compensated by superior optical equipment on the Japanese side.

The Mark 8 fire-control radar had a range accuracy of circa 35-40 yds (32-36 m) at normal battle range (20-25 kyd/18.3-22.9 km), while the director rangefinders of the battleship, having a base length of 26.5 ft (8.08 m) and a magnifying power of 25, had a unit of error of 105-165 yds (96-151 m) at this range.

On the Japanese side, the Type 22 mod. 4 radar had a range accuracy of circa 109 yds (100 m) under optimal conditions. Postulating that the magnifying power of the Japanese rangefinders was the same as those of the American's, then the unit of error of the 49.21 ft (15.0 m) rangefinders was 57-89 yds (52-81 m) at this range, but this was further diminished by that the value fed into the computer was the mean of the measurements of the triplex rangefinders.

Considering that the danger space at this range was about 55-70 yds (50-64 m) in case of a battleship-sized target (about 50 yds/46 m more in case of Yamato's guns, assuming 100 cal. dive), whilst salvo spread was already circa 250-300 yds (229-274 m), the accuracy at which range could be measured was quite satisfactory on both sides.

Table 2.2

Fire-Control Equipment Comparison		
Designation	Iowa	Yamato
Main Battery Rangefinders – Base Length in ft (m)/Type)		
Foretop Director	26.5 (8.08)/S	49.21 (15.0)/Triplex, 2C+1S
Aft Director	26.5 (8.08)/S	32.81 (10.0)/Triplex, 2C+1S
Turret No. I.	46.0 (14.02)/C*	49.21 (15.0)/Triplex, 2C+1S
Turret No. II.	46.0 (14.02)/S	49.21 (15.0)/Triplex, 2C+1S
Turret No. III.	46.0 (14.02)/S	49.21 (15.0)/Triplex, 2C+1S
Main Battery Gunnery Radar)		
Equipment	Mark 8 Gunnery Radar	Type 22 mod. 4 Surface Search/Gunnery Radar
Range (battleship)	40.0 kyd (36.6 km)	38.3 kyd (35.0 km)
Wavelength	10 cm	10 cm
Peak Power	15 kW	2 kW
PRF	1,500	2,500
Pulse Length	0.4 µs	10 µs
Range Accuracy	±15 yds (14 m)+0.1% of Range	±109 yds (100 m) at Optimum Battle Range
Bearing Accuracy	±2 mils	±2-3 deg.
Antenna	"Hair Comb"	"Horn"
Other Equipment)		
Searchlights	36 in (91.4 cm)	59.1 in (150 cm)
Aircraft	3 x Vought OS2U Kingfisher**	7 x Mitsubishi F1M2 Type 0
Computer	Mark 8 Rangekeeper	Type 98 Shagekiban
RPC)		
Training	Yes	No
Elevation	Yes	No

*=Removed
**=Curtiss SC Seahawk from 1945.

Figure 2.1

Mark 38 Director
Mark 8 Radar

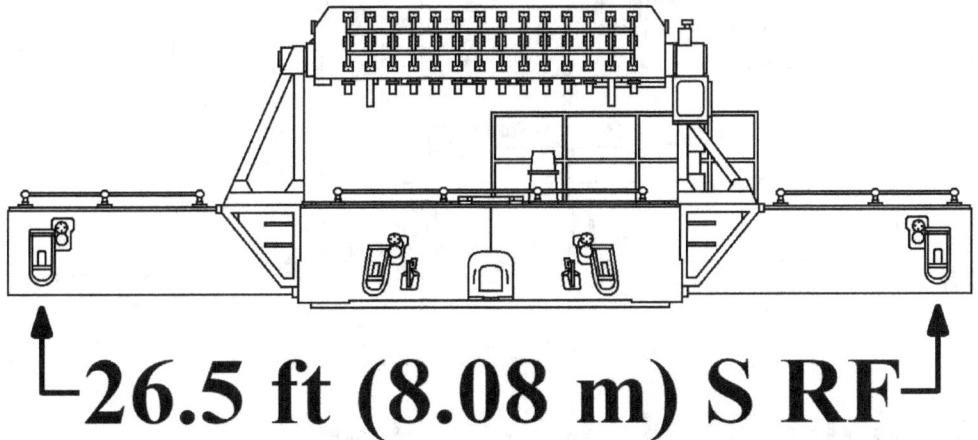

26.5 ft (8.08 m) S RF

Mark 40 Director
Mark 3 Radar

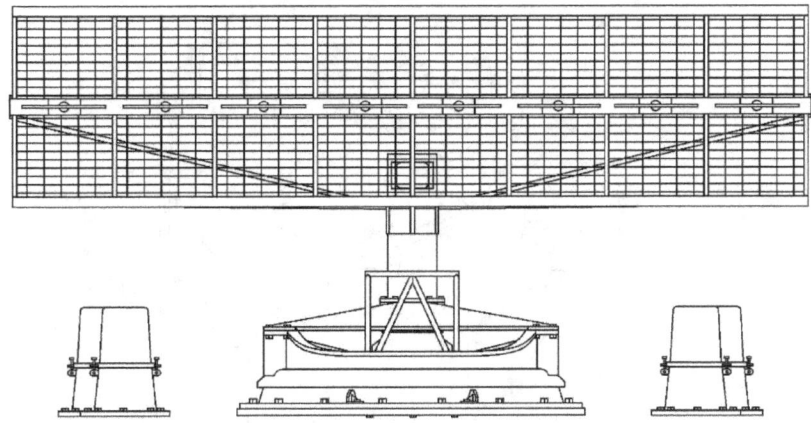

Main Turret

46 ft (14.02 m) S RF

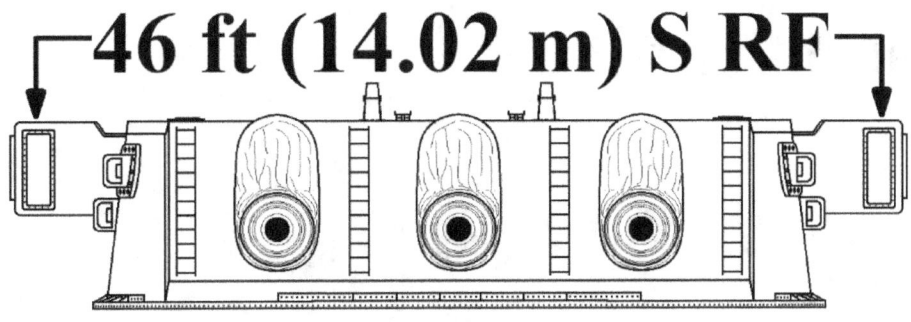

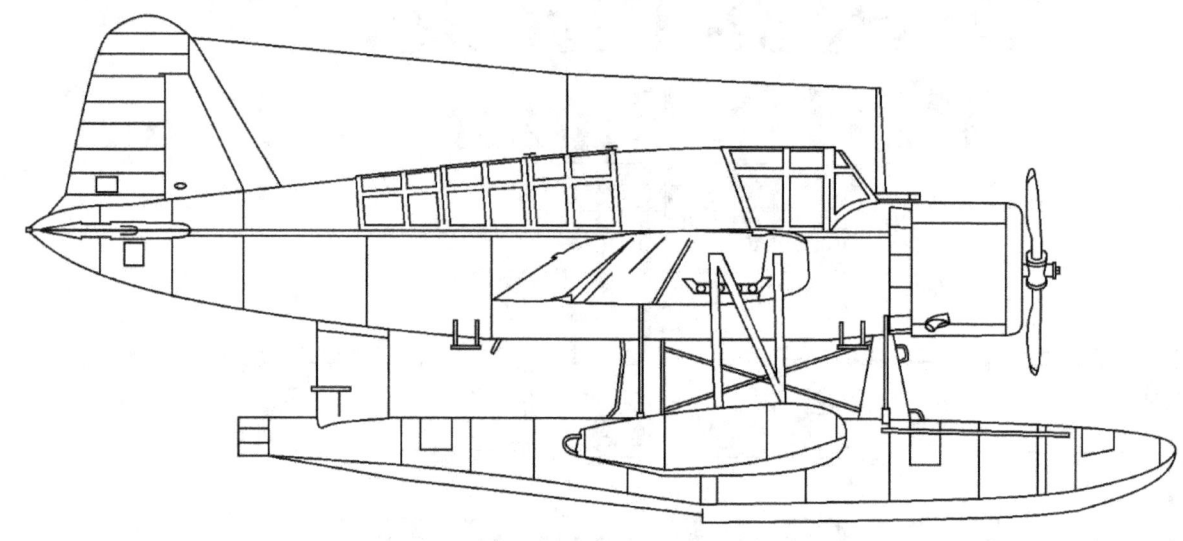

Vought OS2U Kingfisher

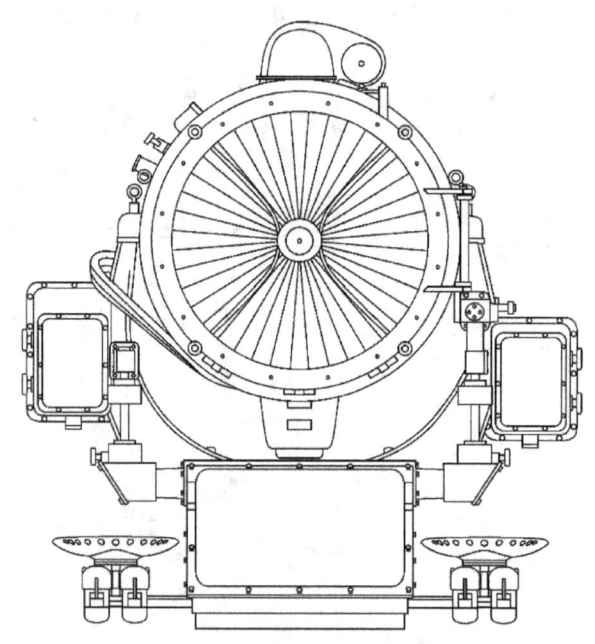

36 in (91.4 cm) Searchlight

Mark 38 GFCS (Block Diagram)

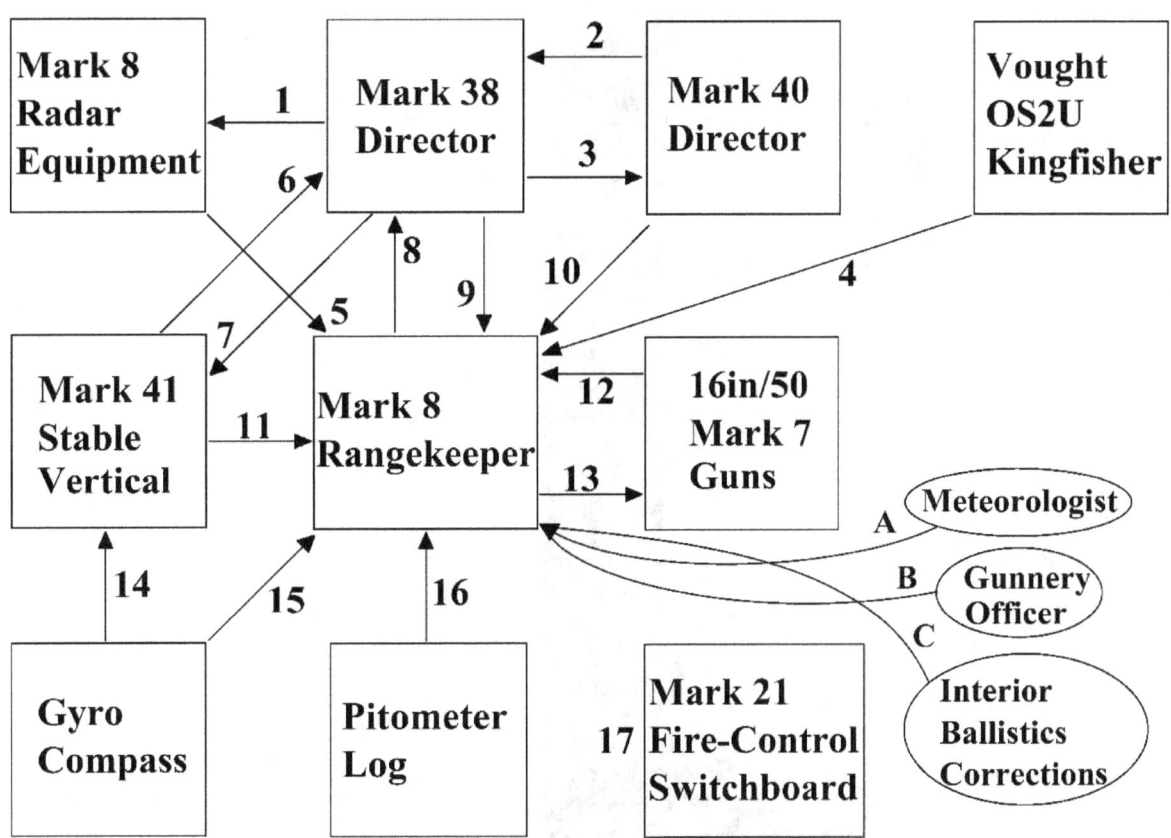

1 Radar Signal (Radar Ranging)
2 Director Train Designation
3 Director Train Designation
4 Range & Deflection Spot
5 Range (Radar Ranging)
6 Level/Cross Level
7 Level/Cross Level
8 Increment of Generated Director Train
 /Generated Director Train/Parallax Range
9 Present Range/Target Bearing & Course & Speed
 /Observed Dirctor Train/Range & Deflection Spot
10 Range & Deflection Spot
11 Level/Cross Level
12 Optical Range Finding
13 Turret Train & Elevation Order
 /Sight Angle & Deflection/Parallax Range
14 Own Ship Course
15 Own Ship Course
16 Own Ship Speed
17 Controls Circuits
A Wind Direction & Speed
B Projectile & Propellant Data
C Initial Velocity

Figure 2.2

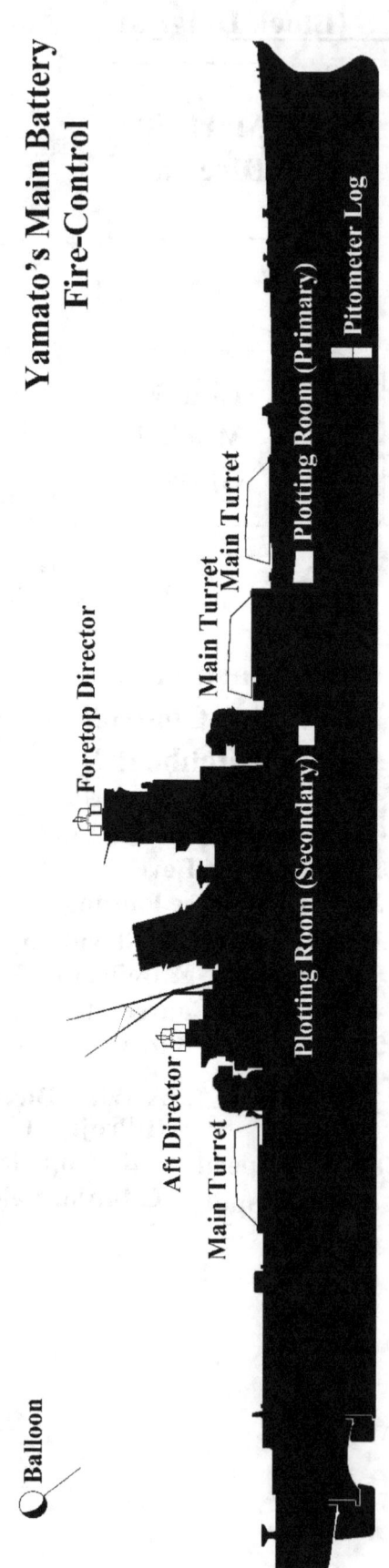

Foretop Director
49.21 ft (15.0 m) Triplex RF (2C+1S)

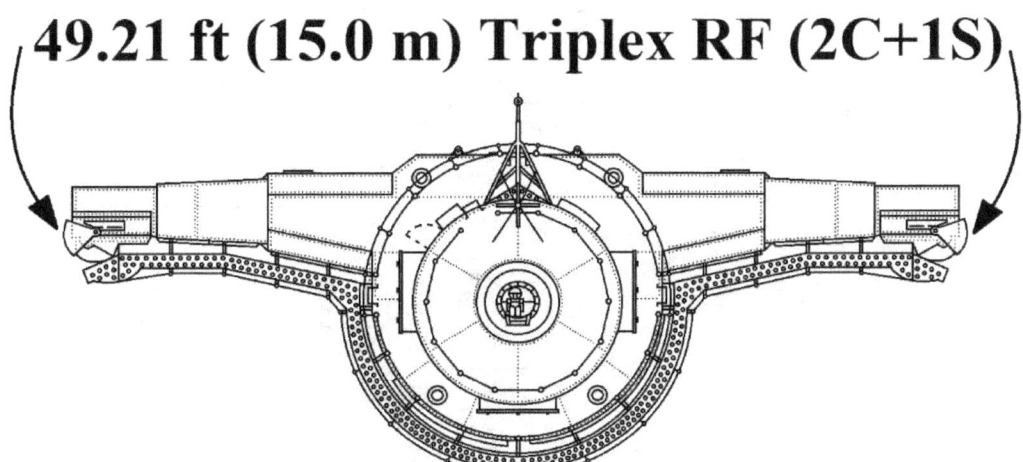

Aft Director
32.81 ft (10.0 m)
Triplex RF (2C+1S)

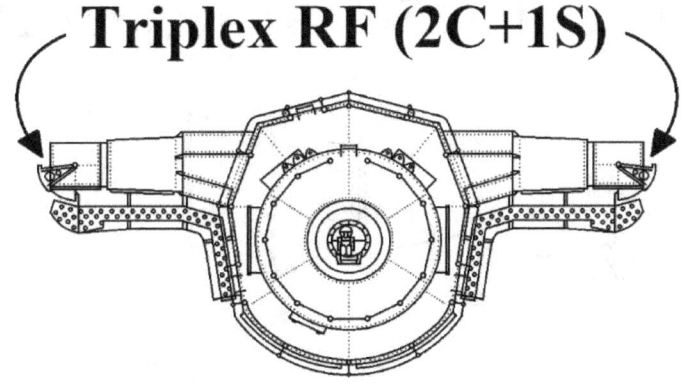

Main Turret

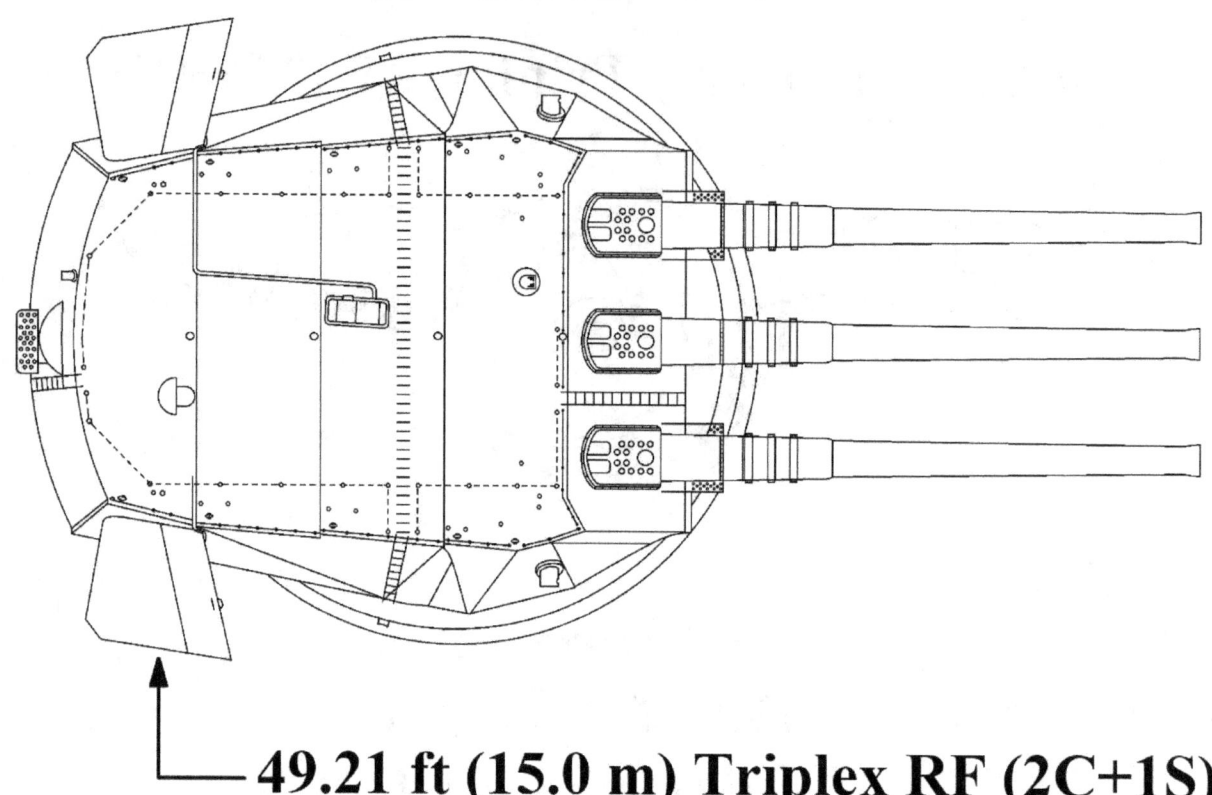

49.21 ft (15.0 m) Triplex RF (2C+1S)

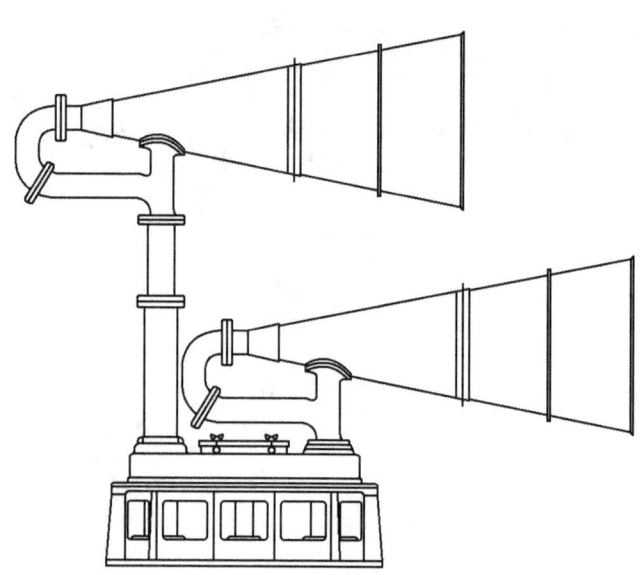

Type 22 mod. 4 Radar

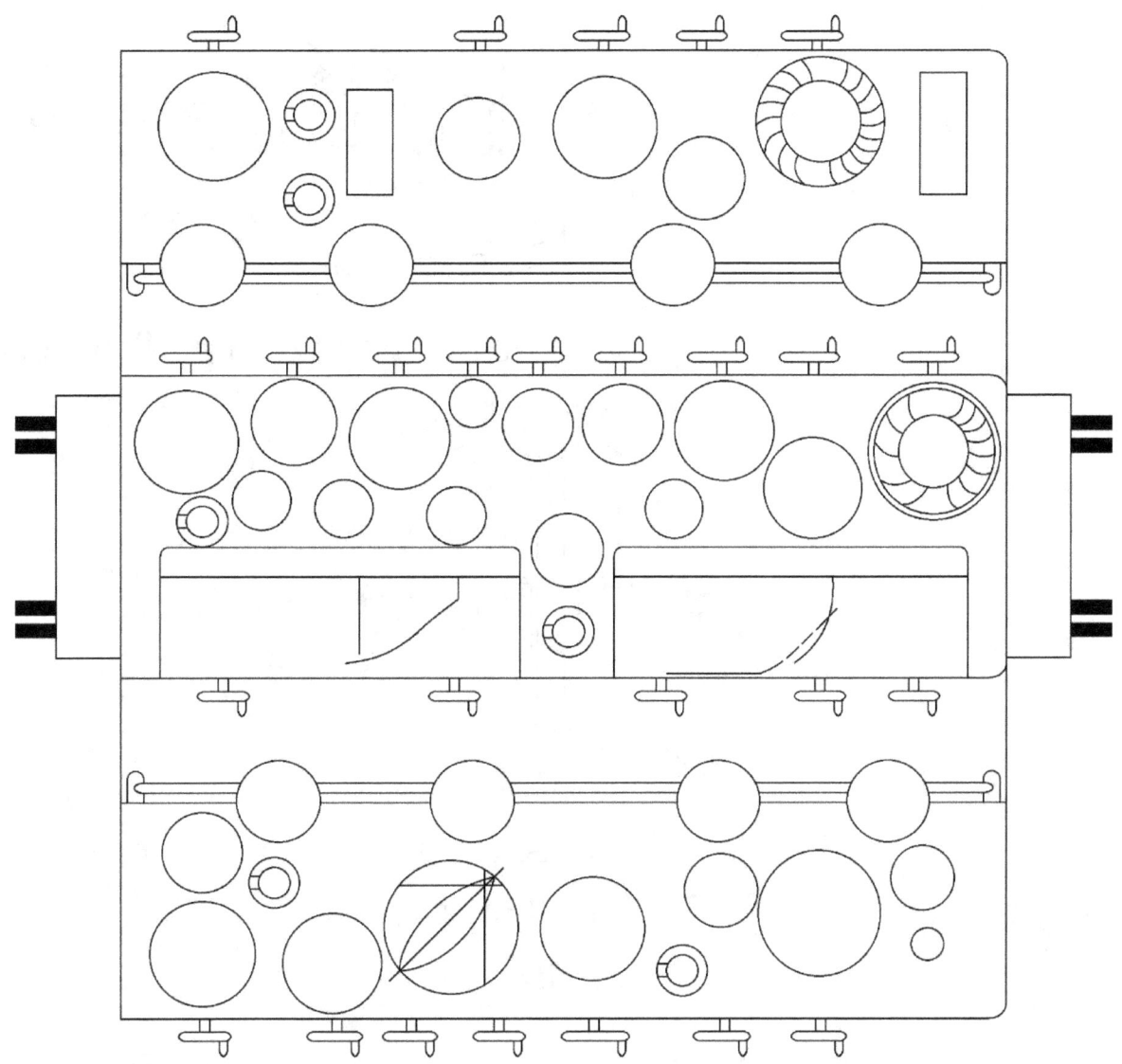

Type 98 Computer

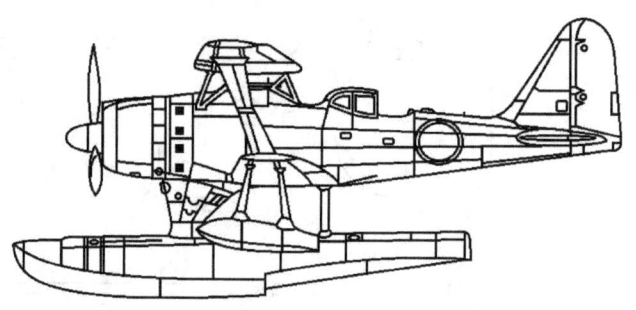

Mitsubishi F1M2 Type 0 Aircraft

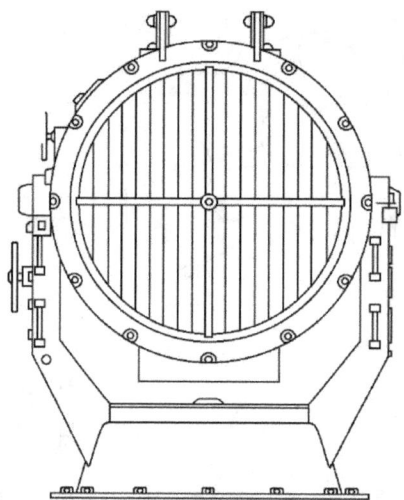

59.1in (150 cm) Searchlight

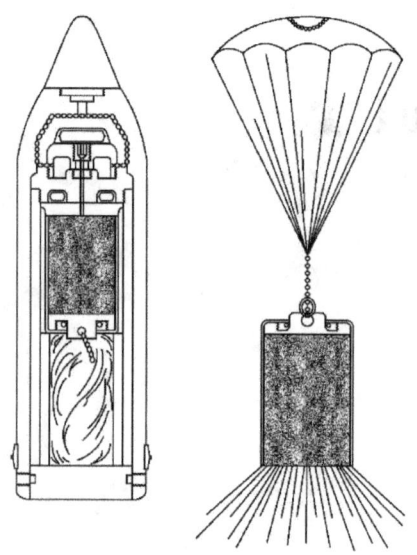

6.1in (155 mm) Starshell

Figure 2.3

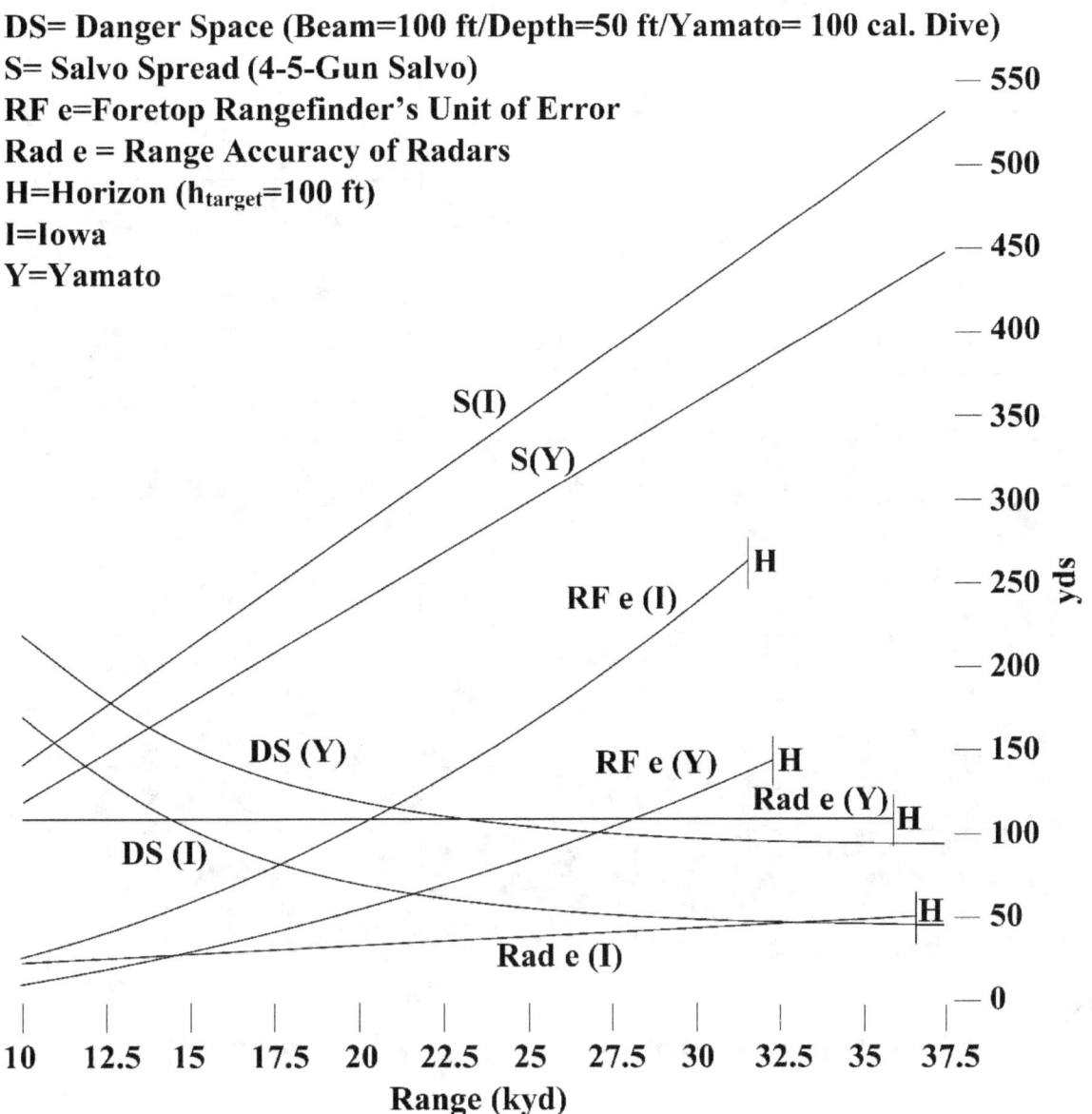

Figure 2.4

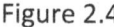

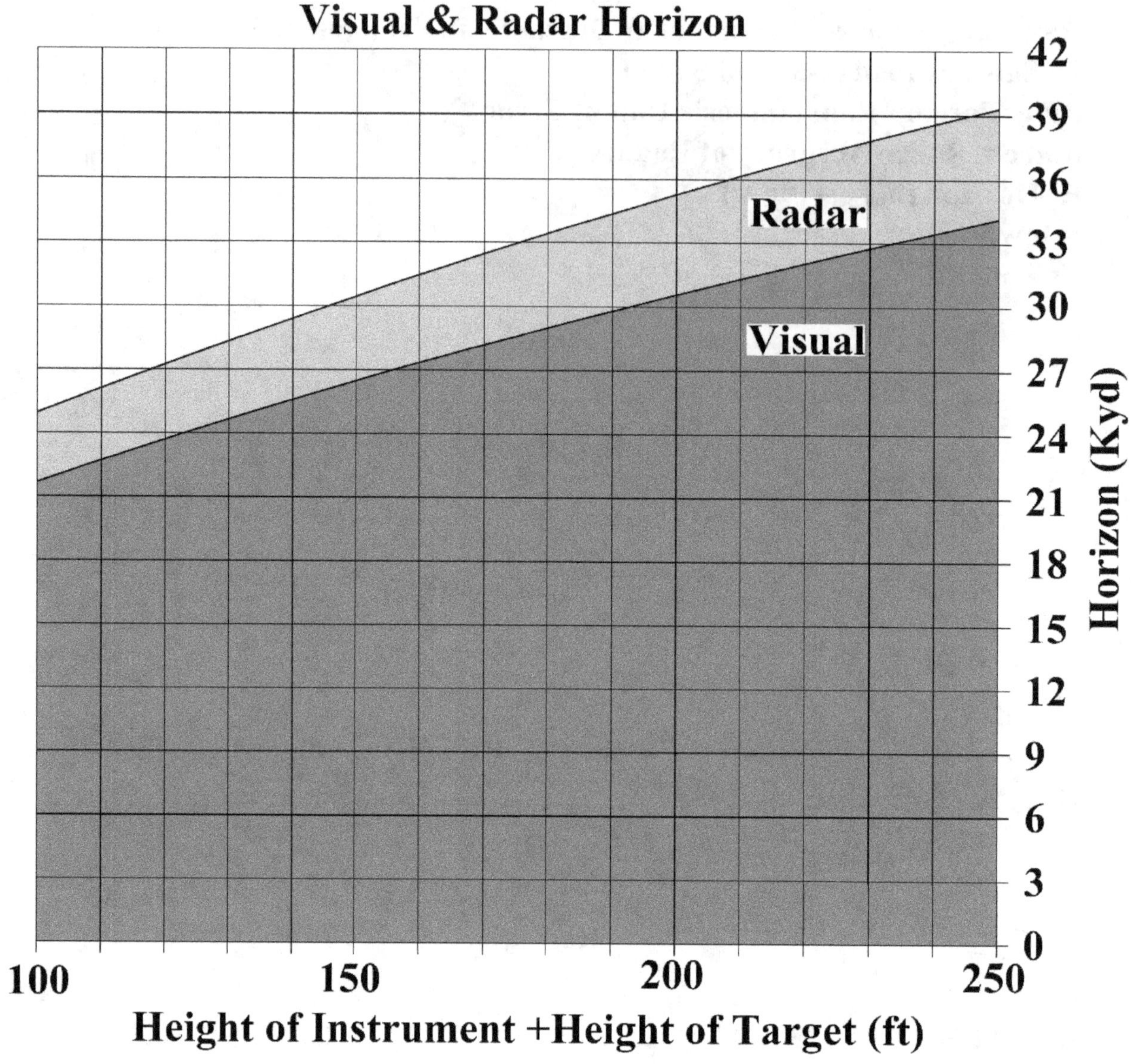
Visual & Radar Horizon
Radar
Visual
42
39
36
33
30
27
24
21
18
15
12
9
6
3
0
Horizon (Kyd)
100
150
200
250
Height of Instrument +Height of Target (ft)

Chapter 3 – Protection

Iowa's Protection Scheme

The first modern battleship class commissioned by this nation – North Carolina – was designed in an era of volatile international restrictions on capital ship armament and displacement. Even though main gun diameter was limited at 14″ (356 mm) by the Second London Naval Treaty, American negotiators urged to include an "escalator clause" in case any of the major navies should remain reluctant to ratify the agreement by 1 April 1937. Despite the fact that the Americans succeeded in simultaneously designing quadruple 14″ (356 mm) and triple 16″ (406 mm) mounts that were interchangeable – hence could invoke the "escalator clause" and mount the larger guns on the new ships – the protection scheme of North Carolina was designed against 14″ (356 mm) attack.

When the succeeding South Dakota class was designed, the primary objective of the designers was to secure the ship against 16″ (406 mm) attack without having to raise her displacement as compared to the previous ships. The following modifications were made to reach this aim:
  - The length of the ship and the relative length of her citadel were decreased.
  - The main belt was moved inside the hull and its inclination was amplified.
  - A more concentrated horizontal protection scheme was proposed.
  - Side armour abaft of the citadel was inclined from the vertical.
  - The forward arcs of the main turret barbettes were more intensively tapered.

With these modifications, the designers managed to provide a decent period of immunity to South Dakota against 2,240 lbs (1,016 kg) 16″ (406 mm) projectiles with an initial velocity of 2,520 fps (768 mps). However, immunity against 2,700 lbs (1,225 kg) shells of the same diameter was marginal at best.

Iowa's protection scheme was based on that of the preceding South Dakota class, though a few minor alterations were made. While the new ship carried palpably heavier armour, this was largely the result of her longer hull necessitating a larger citadel to attain the same relative protected length; armour distribution, plate thicknesses and immunity zones were almost the same as in the South Dakota design, albeit the new ship had a very slight edge.

The Iowa class battleships had a main deck of adequate thickness to initiate the fuze of aerial bombs, but most armour reserved for horizontal protection was concentrated into a heavy, laminated armour deck at the second deck level. This heavy armour deck was followed by two lightly armoured decks abreast the ship's machinery compartments, namely a splinter deck and the third deck. There was no splinter deck above the magazines, but the thickness of the third deck was slightly boosted to compensate for this.

The upper edge of the main armour belt was connected to the sides of the second deck. With the intention of enhancing its effective resistance, it was inclined from the vertical. The thickness of the hull was increased in front of the main belt. The thickened hull plating protected the compartments outside the main belt against shrapnel, and, in theory, could knock off the cap of incoming AP shells, albeit neither its thickness (1.5″/38 mm) nor inclination (vertical) was suitable to make it a reliable decapping plate against major caliber shells. (The Italians used 2.76″+.35″/70+9 mm plates inclined 15 deg. from the vertical for this purpose.)

The main armour belt was relatively shallow, but it was followed by an excessively deep tapering lower belt. This was provided as an attempt to secure the ship against underwater hits. Unfortunately, the torpedo defense system, which was formed by the lower belt and three lighter longitudinal bulkheads, two between the lower belt and the hull, and one behind the former, was made rigid by the heavy ballistic armour carried below the waterline. The outer two of the four longitudinal compartments forming the ship's side protection system were liquid loaded, but the inner two were kept void.

The citadel was enclosed by vertical transverse bulkheads. In the last two units of the class, the thickness of these bulkheads was increased. The lower strake of the forward transverse bulkhead was tapered, whilst the aft transverse bulkhead was virtually unarmored below the third deck level. The former design trait was justified by the fact that transverse bulkheads were unlikely to be hit at a right angle, hence their lower strake could only be reached by projectiles that previously travelled considerable distances submerged, losing much of their velocity, while the latter by the shielding effect of heavy armour extended abaft the citadel.

As a means to protect the steering gear machinery and shaft tunnels, the third deck was heavily armoured abaft the citadel, and its edges were connected to a heavy tapering lower belt. This system ensured comparable ballistic resistance as the citadel. The system was terminated by a transverse bulkhead abaft of the steering gear machinery.

No heavy armour was carried outside the forward transverse bulkhead, but the torpedo bulkheads were extended to strengthen the bow section.

The main turrets, turret barbettes and conning position were protected by the heaviest armour plates carried by the ship. This was necessary to match the effective resistance of the inclined armour belt and the three to four armoured decks forming the citadel.

The secondary guns and directors had extensive splinter protection but were not protected against direct hits.

Heavy vertical armour plates above the waterline were constructed of Class A type armour, except for the turret face and side plates, and the side armour of the conning position. These, the lower armour belt, and the heavy horizontal armour of the ship, were made of Class B type armour. Light armour plates were made of STS, with the exception of the underwater hull and the torpedo bulkheads, which were constructed of HTS.

Based on ADM documents, the British regarded the effective resistance of heavy U.S. armour comparable to that of their own. However, U.S. NPG reports indicate that German cemented armour was narrowly superior to their own, even though the British considered their cemented armour superior to that of German manufacture. Ballistic tests conducted in the U.S. infer that American non-cemented armour was moderately superior to the German equivalent, albeit this too was the reversal of the British findings, which hint that German non-cemented armour was of very high quality. Based on post-war ballistic tests, the Americans appraised their armour of higher quality as compared to Japanese armour. However, ADM documents imply the opposite, namely that heavy Japanese armour was superior to both British and German materials.

Picture 3.1

17.3 in (439 mm) side armour of New Jersey's conning tower. The picture was taken on 1 July 1981.

Yamato's Protection Scheme

Yamato's protection scheme was completely different from those of the preceding Nagato class and interwar period battleship designs. As far as ballistic protection is concerned, the designer's primary intent was to secure the ship against the new 18.11" (460 mm) guns at a range of 21,872-32,808 yds (20,000-30,000 m) at 90 deg. target angle. Owing to the magnitude of the threat specified, this was a very challenging task. As if the overwhelming piercing capacity of the new caliber was not difficult enough to deal with, the designers also had to take the enhanced underwater performance of the new shells into account, which necessitated very heavy ballistic protection below the waterline.

The Japanese adopted practically all available weight-saving methods to meet design specifications. The new ships had an "all-or-nothing" type protection scheme, concentrated horizontal armour and armament, as well as a compact citadel. All side armour was inclined from the vertical. The only heavily armoured regions outside the citadel were the ship's main and auxiliary steering gear machinery compartments.

Yamato had a lightly armoured flying deck constructed of DS atop her machinery and CNC abreast her magazines and outside the citadel. Atop the machinery, this deck was followed by a light DS upper deck, but this terminated before reaching the forward magazines. This was compensated by the greater thickness of the CNC flying deck in this region. The main protective deck was at middle deck level. Its sides had increased thickness and were slightly sloped from the horizontal. The main armour deck was followed by a light splinter deck and a light lower deck.

The upper edge of the main belt was attached to the sides of the main armour deck. The main belt was deep and inclined from the vertical. Abreast the machinery compartments, it was followed by a tapering lower belt of reduced inclination.

The lower belt also functioned as the main holding bulkhead, despite the possibility that, as in contemporary U.S. designs, this made the torpedo defense system rigid. It was followed by two additional holding bulkheads. These, combined with the blistered hull in front of the lower belt, formed the torpedo defense system of the battleship. The Japanese preferred to keep all compartments void.

Abreast the magazines, both the thickness and the inclination of the lower belt were increased. Furthermore, the lower edge of the port and starboard side lower belt strakes were connected by a magazine bottom deck of appreciable thickness, protecting the magazines from underwater detonations. The magazine bottom deck was terminated by vertical bulkheads at the limits of the magazines and machinery compartments.

The citadel was enclosed by a combination of diagonal and transverse bulkheads. The former type was introduced as an attempt to save weight. Further weight was conserved by inclining both types of bulkheads from the vertical. The upper edges of all bulkheads were connected to the main armour deck by inclined crowns, while their lower edges marked the limits of the ship's magazine bottom deck.

The Japanese placed high emphasis on protecting the uptakes of their newest battleships. A very heavy armour grating, nearly twice the thickness of the main deck, was provided. This was shielded by the armour plating the funnel was wrapped in.

The main steering gear compartment was protected by a small but very heavy armour box. Like the citadel, this too was octagonal in the horizontal plane and had inclined sides. The effective resistance of this system closely matched that of the citadel. With the aim of further securing steering ability, Yamato was furnished with a secondary steering gear machinery, situated sufficiently far from the main steering gear engine room that a single torpedo hit could not disable them both. The secondary steering gear room had a similar protection scheme, but slightly lighter side armour than the main. This was justified by the fact that it was deeper inside the ship; therefore, diving projectiles reaching it necessarily had to have reduced terminal velocity.

The main turrets, turret barbettes, and the conning position had extraordinarily heavy armour. Even against the new 18.11" (460 mm) guns, the main turrets were, in theory, invulnerable from a point-blank range up to extremely long ranges. The thickness of the beams of the barbettes and that of the conning position was appropriate to equal the effective resistance of the inclined main belt under the fighting conditions envisaged.

The secondary turrets had armour plating adequate against shrapnel, but not direct hits.

The directors had bulletproof plating, designed to withstand 0.5" (12.7 mm) strafing fire at a range of 219 yds (200 m).

Heavy vertical armour above the waterline was constructed of VH, heavy horizontal armour of MNC. Heavy vertical armour plates below the waterline were made of NVNC. Light armour plates were made of CNC. Torpedo bulkheads, protective plating, and the backing of heavy armour plates were made of DS.

Based on U.S. NPG reports, Japanese armour was slightly inferior to most foreign types. However, based on ATC documents, heavy Japanese armour had higher effective resistance than any other foreign type.

Picture 3.2

Yamato running full-power trials in Sukumo Bay, 30 October 1941. Notice the upper edge of the main belt of the ship.

Comparison of Protection Schemes

The distribution of heavy armour plates was remarkably similar in the two designs, except for the following:

1) Yamato had a relatively heavy magazine bottom deck, which Iowa lacked.

2) Yamato's main armour belt was deeper, i.e. side armour began to taper deeper below the waterline.

3) Iowa's citadel was rectangular; that of Yamato was octagonal.

4) Yamato's bulkheads were inclined from the vertical; Iowa's were vertical.

5) Iowa had an extensive protective system abaft her citadel. In the Japanese design, only local armour protection was provided for the two steering gear rooms.

6) Inclination of armour plates (belts, turret faces, bulkheads) was greater in the Japanese design, except for the lower belt abreast the machinery compartments.

7) Iowa's armour belt was internal and shielded by hull plating; Yamato's was external.

8) Yamato carried heavier armour; her protection scheme was designed to yield immunity against a significantly more powerful gun.

As can be seen, even though the American ship had much better ballistic underwater protection and more concentrated armour scheme than her European contemporaries, the Japanese placed an even higher emphasis on underwater protection and weight saving. The protection of the primary armament, conning position and steering ability of both ships compared very favorably with European designs too, albeit their torpedo protection was most likely inferior by reason of the rigidity of the heavy underwater armour plating and the relatively unremarkable depth of the side protection system characterizing both the American and the Japanese designs.

Table 3.1

Designation	Thickness		Material
	in	mm	
Iowa – Protection			
Citadel)			
Hull	0.625-1.5-0.625	16-38-16	STS-STS-HTS
Main Belt (19°)	12.1+0.875	307+22	Class A+STS
Lower Belt (19°)	12.1-1.625+0.875	307-41+22	Class B+STS
Main Deck	1.5	38	STS
Second Deck	4.75+1.25	121+32	Class B+STS
Splinter Deck (Machinery only)	0.625	16	STS
Third Deck – Machinery	0.5-0.625	13-16	STS
Third Deck – Magazines	1	25	STS
Magazine Bulkheads	1.5	38	STS
Forward Transverse Bulkhead (BB 61 & 62)	0.625-11.3-8.5	16-287-216	STS-Class A
Aft Transverse Bulkhead (BB 61 & 62)	0.625-11.3-0.625-2	16-287-16-51	STS-Class A-STS
Forward Transverse Bulkhead (BB 63 & 64)	0.625-14.5-11.7	16-368-297	STS-Class A
Aft Transverse Bulkhead (BB 63 & 64)	0.625-14.5-0.625-2	16-368-16-51	STS-Class A-STS
Torpedo/Splinter Bulkheads	0.625	16	HTS/STS
Main Turrets)			
Face Plate (36°10'18")	17+2.5	432+64	Class B+STS
Side Plate	9.5+0.75	241+19	Class B+STS
Rear	12	305	Class A
Roof	7.25	184	Class B
Barbettes)			
Beam	17.3	439	Class A
Forward & Aft	14.8-11.6	376-295	Class A
Lower	3-1.5	76-38	STS
Conning Tower)			
Sides	17.3	439	Class B
Forward	17.3	439	Class B
Aft	17.3	439	Class B
Roof	7.25	184	Class B
Deck	4	102	STS
Tube	16	406	Class B
Steering Gear Room)			
Sides (19°)	13.5	343	Class A
Aft Bulkhead	11.3	287	Class A
Fwd. Bulkhead	0.625	16	STS
Roof	6.2	157	Class B
Deck	1.5	38	STS
Shaft Tunnels)			
Sides (19°)	13.5-7.125-5.625	343-181-143	Class B
Deck	5.6+0.75	142+19	Class B+STS
Secondary Mounts)			
Face Plate	2.5	64	STS
Sides	2.5	64	STS
Rear	2.5	64	STS
Casemates	2.5	64	STS
Superstructure)			
Directors	1.5	38	STS
Director Tubes	1-2.5	25-64	STS
Fire-Control Towers	0.5-1	13-25	STS

Table 3.2

Yamato – Protection			
Designation	Thickness		Material
	in	mm	
Citadel)			
Upper Side Hull between Flying Deck and Upper Deck	0.98+0.87	25+22	DS+DS
Upper Side Hull between Upper Deck and Middle Deck	0.98	25	DS
Main Belt (20°)	16.14+0.63	410+16	VH+DS
Lower Belt – Machinery (14°)	7.87-1.97+0.55	200-50 +14	NVNC+DS
Lower Belt – Magazines (25°)	10.63-3.94	270-100	NVNC
Flying Deck – Midship/Center	0.47	12	DS
Flying Deck – Midship/Sides	0.79+0.71	20+18	DS+DS
Flying Deck – Magazines/Center	1.97	50	CNC
Flying Deck – Magazines/Sides	1.38	35	CNC
Upper Deck – Midship/Center	0.39	10	DS
Upper Deck – Midship/Sides	0.98	25	DS
Middle Deck – Center	7.87+0.39	200+10	MNC+DS
Middle Deck – Sides (7°)	9.06+0.55	230+14	MNC+DS
Splinter Deck	0.35	9	DS
Lower Deck	0.35	9	DS
Forward Transverse Bulkhead (25°)	11.81	300	VH
Aft Transverse Bulkhead – Upper (25°)	11.81	300	VH
Aft Transverse Bulkhead – Lower (25°)	10.63	270	VH
Diagonal Bulkhead – Crowns (25°)	13.39	340	MNC
Diagonal Bulkhead – Forward (25°)	13.78	350	VH
Diagonal Bulkhead – Aft (25°)	12.99-13.78	330-350	VH
Torpedo/Splinter Bulkheads	0.35-0.63	9-16	DS
Magazine Bottom Deck	1.97-3.15	50-80	CNC
Magazine Bottom Deck Limiting Bulkhead	1.97-2.95-3.94	50-75-100	NVNC
Main Turrets)			
Face (45°)	25.98	660	VH
Fore Sides	12.99	330	VH
Rear Sides	9.84	250	VH
Roof	10.63	270	MNC
Rear	7.48	190	VH
Overhang	1.97	50	---
Barbettes)			
Beam	22.05	560	VH
Forward & Aft	17.32-14.96	440-380	VH
Lower	1.97	50	CNC
Conning Tower)			
Beam	22.05	560	VH
Forward	19.69	500	VH
Aft	15.75	400	VH
Roof	7.87	200	MNC
Tube	11.81	300	VH
Deck	2.95	75	CNC
Steering Gear Room)			
Sides (25°)	14.17	360	VC
Forward (25°)	14.17	360	VC
Aft (25°)	13.78	350	VC
Roof	7.87	200	MNC
Deck	0.98	25	DS
Auxiliary Steering Gear Room)			
Sides (25°)	11.81	300	VH
Forward	11.81	300	VH
Aft	9.84	250	VH

Roof	7.87	200	MNC
Deck	0.79	20	DS
Secondary Turrets)			
Face	1.97	50	CNC
Sides	1.97	50	CNC
Rear	1.97	50	CNC
Roof	1.97	50	CNC
Barbette	1.97+0.98	50+25	CNC+DS
Uptakes)			
Funnel	1.97	50	CNC
Grating	14.96	380	MNC
Bow and Stern)			
Flying Deck – Center	1.97	50	CNC
Flying Deck – Sides	1.38	35	CNC

Table 3.3

Iowa – Dimensions			
Designation	ft	m	% of Waterline Length
Citadel – Length	464	141.4	54.0
Main Armour Belt – Max. Depth	10.6	3.2	---
Main Armour Belt – Min. Depth	9.9	3.0	---
Lower Armour Belt – Max. Depth	27.5	8.4	---
Lower Armour Belt – Min. Depth	21.4	6.5	---
Armour Belt – Max. Total Depth	38.1	11.6	---
Armour Belt – Min. Total Depth	31.3	9.5	---
Lower Belt Aft – Length	92.0	28.0	10.7
Lower Belt Aft – Max. Depth	17.0	5.2	---
Lower Belt Aft – Min. Depth	16.6	5.1	---
Steering Gear Belt – Length	56	17.1	6.5
Steering Gear Belt – Max. Total Depth	9.8	3.0	---
Side Protection – Max. Depth	17.9	5.5	---

Table 3.4

Yamato – Dimensions			
Designation	ft	m	% of Waterline Length
Protected Length	457.3	139.4	54.5
Main Armour Belt – Depth	18.0	5.5	---
Lower Armour Belt – Depth (Midship)	26.2	8.0	---
Armour Belt – Total Depth (Midship)	44.2	13.5	---
Torpedo Defense System – Depth (Midship)	16.4	5.0	---

Table 3.5

Iowa's Immunity Vs. Yamato's Guns at 90 deg. Target Angle Based on Bu. Ord. Sk. 78841 – yds (m)			
Designation	Inner	Outer	Outer-Inner
Citadel	24,800 (22,677)	29,800 (27,249)	5,000 (4,572)
Turrets	24,700 (22,586)	31,600 (28,895)	6,900 (6,309)
Barbettes	26,500 (24,232)	31,600 (28,895)	5,100 (4,663)
Steering	24,700 (22,586)	28,800 (26,335)	4,100 (3,749)
CT	26,500 (24,232)	31,600 (28,895)	5,100 (4,663)

Table 3.6

Iowa's Immunity Vs. Yamato's Guns at 90 deg. Target Angle Based on the DeMarre Formula (K coefficient adjusted to official data) – yds (m)			
Designation	Inner	Outer	Total
Citadel	29,000 (26,518)	22,200 (20,300)	-6,800 (-6,218)
Turrets	45,965 (42,030)	25,200 (23,043)	-20,765 (-18,988)
Barbettes	31,000 (28,346)	25,200 (23,043)	-5,800 (-5,304)
Steering	29,000 (26,518)	21,000 (19,202)	-8,000 (7,315)
CT	31,000 (28,346)	25,200 (23,043)	-5,800 (-5,304)

Table 3.7

Yamato's Immunity Vs. Iowa's Guns at 90 deg. Target Angle Based on Bu. Ord. Sk. 78841* – yds (m)			
Designation	Inner	Outer	Outer-Inner
Citadel	17,000 (15,544)	34,500 (31,547)	17,500 (16,002)
Turrets	0 (0)**	36,800 (33,650)	36,800 (33,650)
Barbettes	16,600 (15,179)	36,800 (33,650)	20,200 (18,470)
Steering	19,100 (17,465)	33,400 (30,541)	14,300 (13,076)
CT	16,600 (15,179)	33,400 (30,541)	16,800 (15,362)
*=The official penetration curves of Iowa predict the same periods of immunity. **=25.98 in (660 mm)@45° turret face was impenetrable. (See Appendix C).			

Figure 3.1

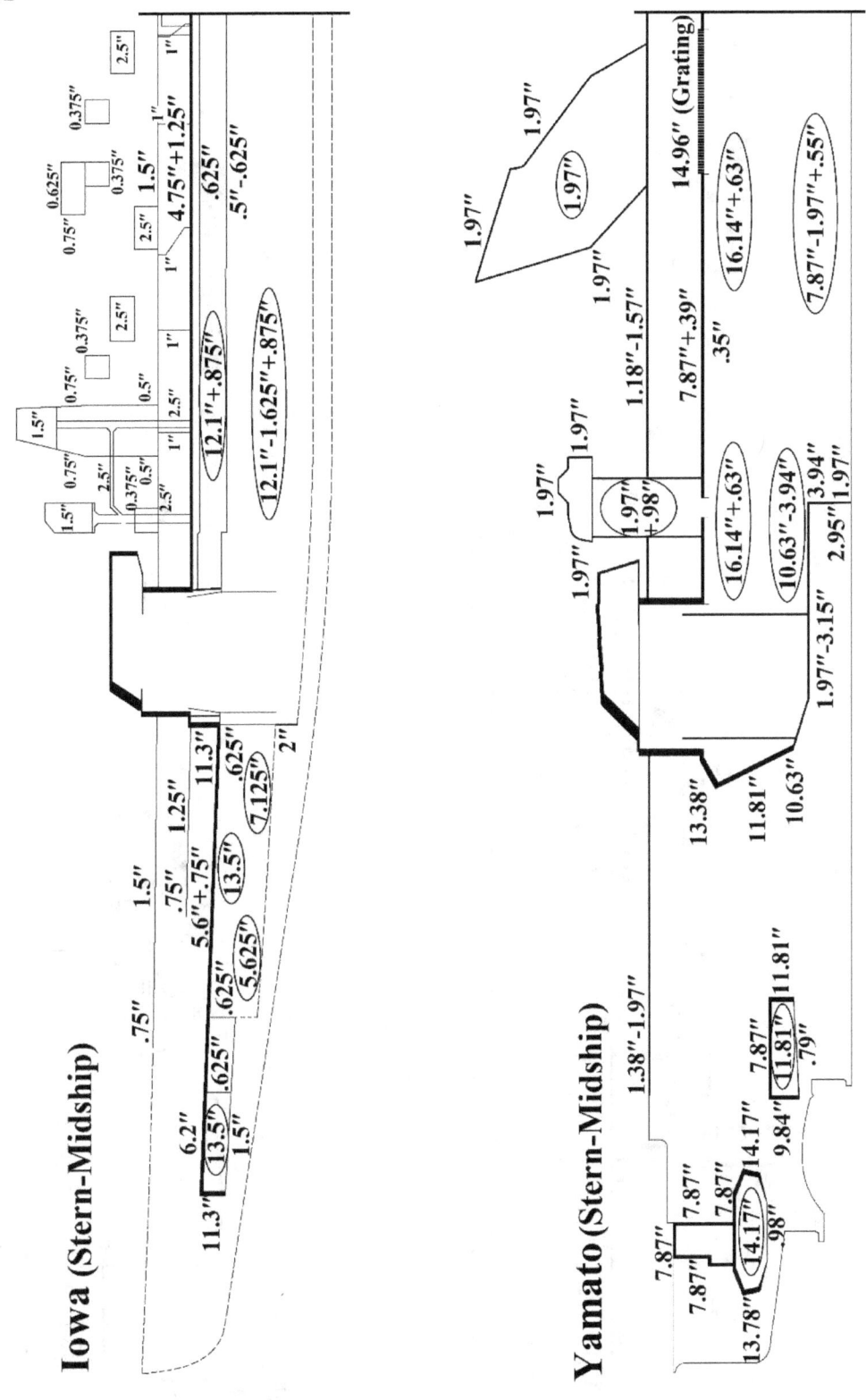

Iowa (Midship-Bow)
.75"
.75"
1.5"
1.25"
17"+2.5"
11.6"
11.3"
8.5"
7.25"
9.5"+.75"
17.3"
3"
1.5"
12"
11.6"
1"
17.3"
7.25"
4"
0.625"
16"
1"
0.375"
1.5"
2.5"
0.625"
0.75"
12.1"+.875"
12.1"-1.625"+.875"
0.375"
2.5"

Yamato (Midship-Bow)
1.38"-1.97"
14.96"
13.38"
11.81"
25.98"
10.63"
12.99"
22.05"
1.97"
9.84"
7.48"
14.96"
1.97"-3.15"
16.14"+.63"
10.63"-3.94"
19.69"
7.87"
3.94"
1.97"
2.95"
15.75"
2.95"
11.81"
22.05"

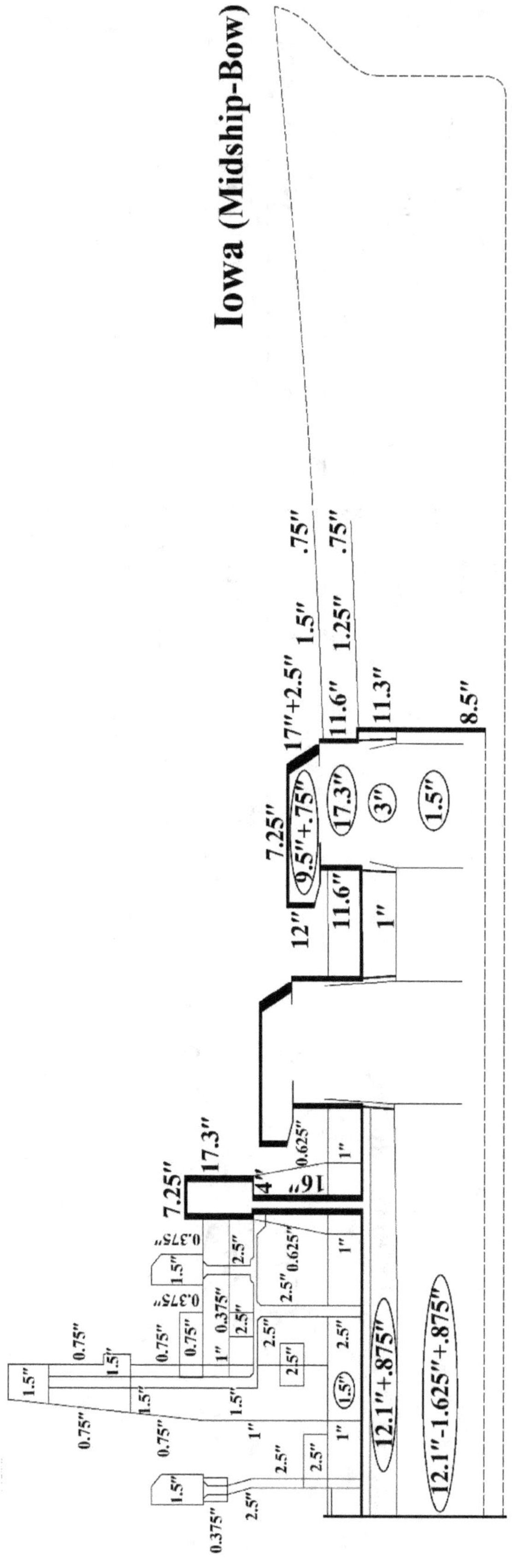

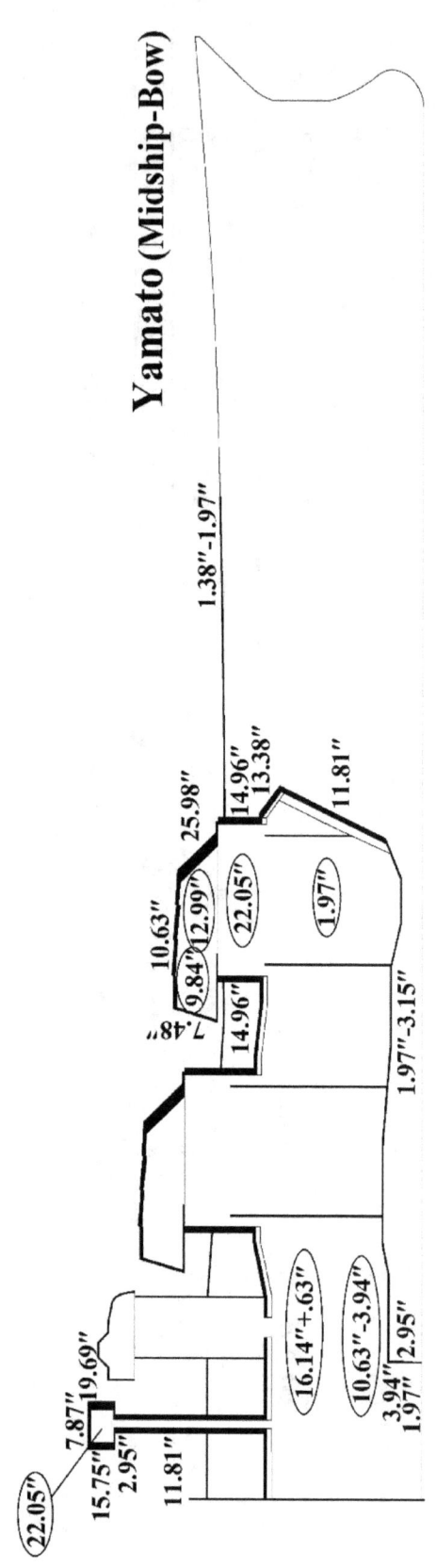

Figure 3.2

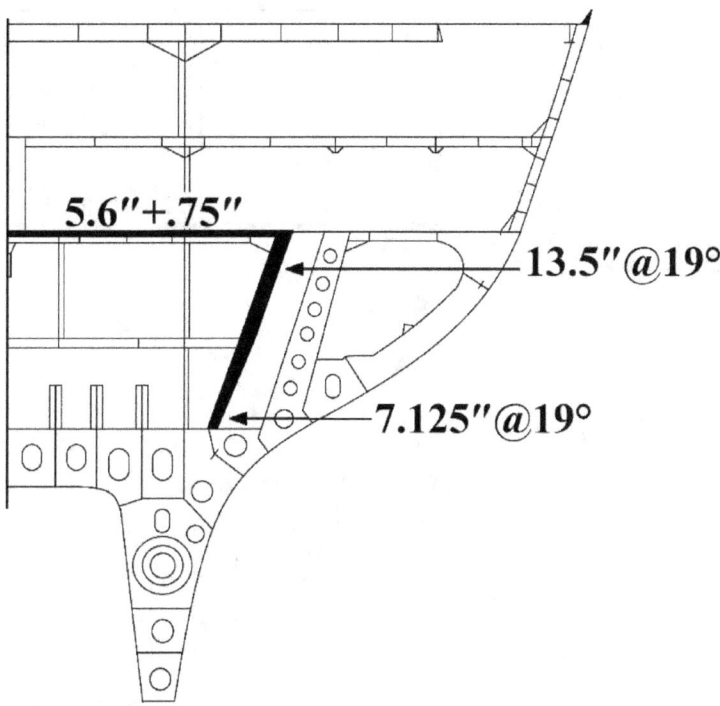

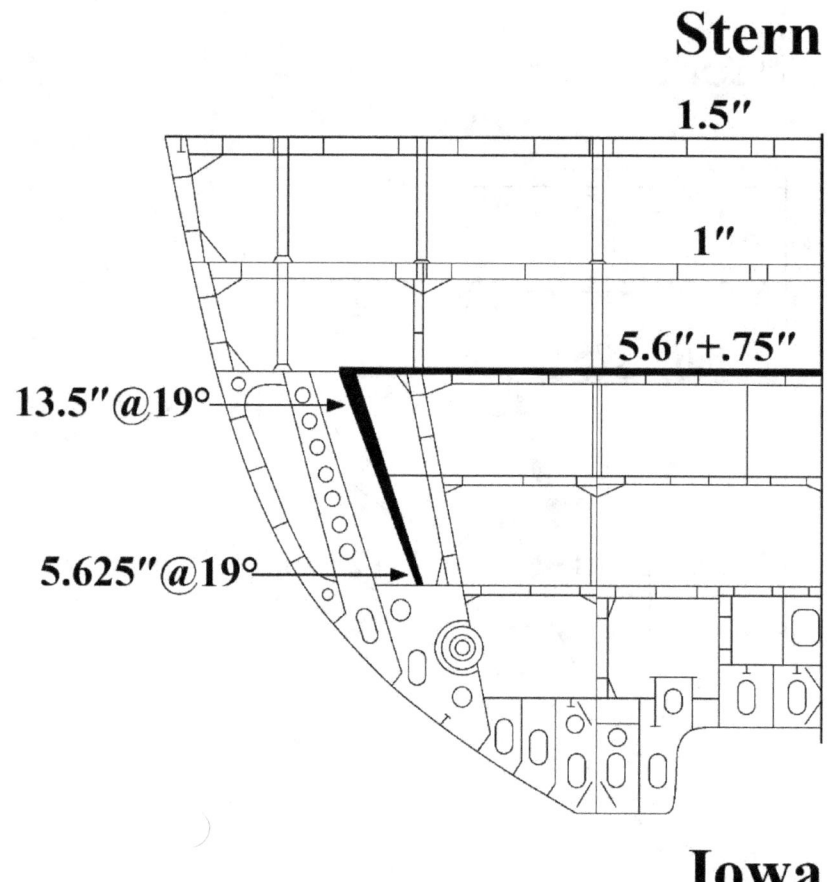

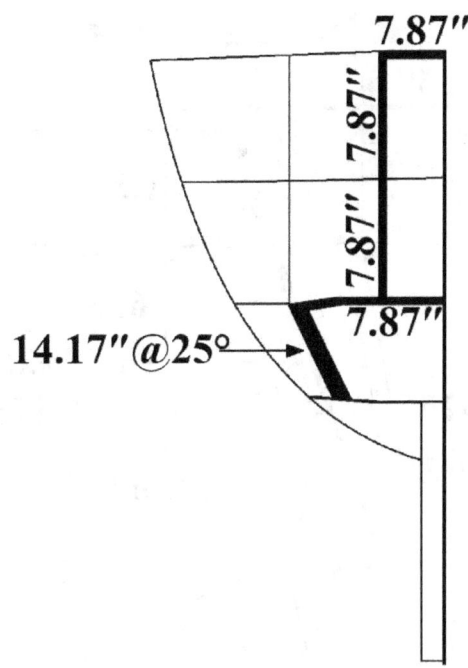

Machinery

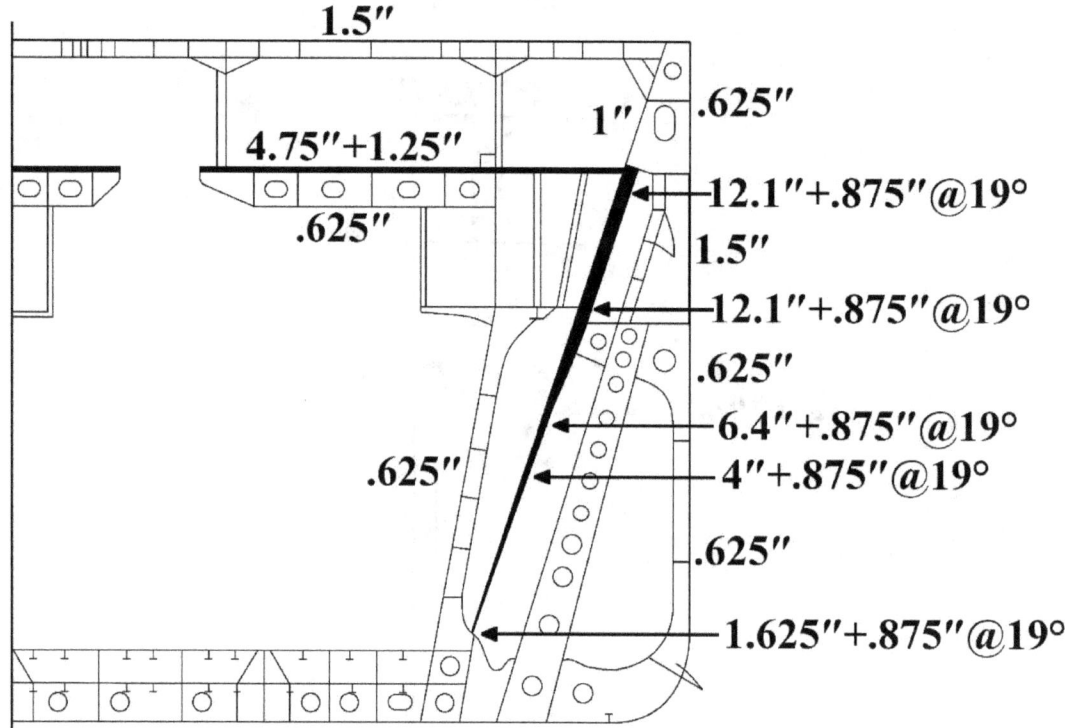

Iowa

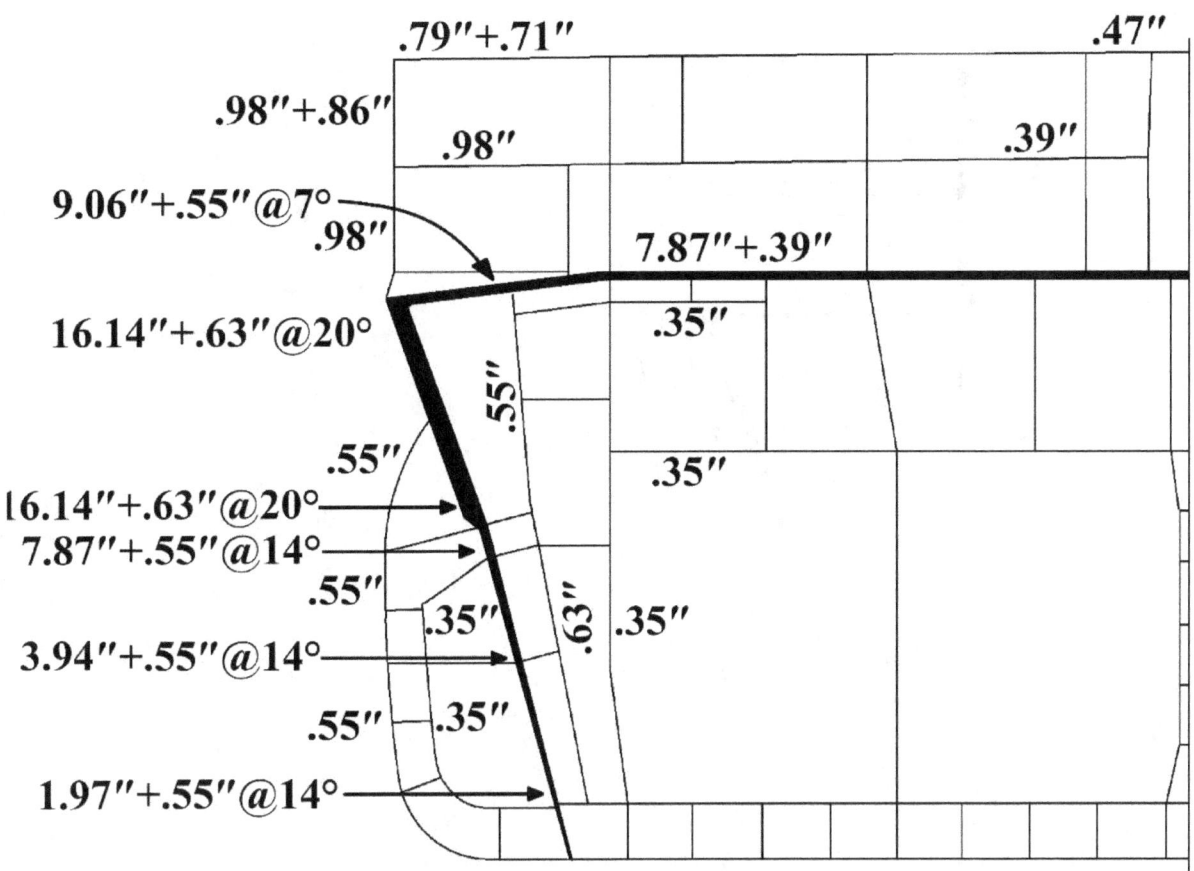

Magazines

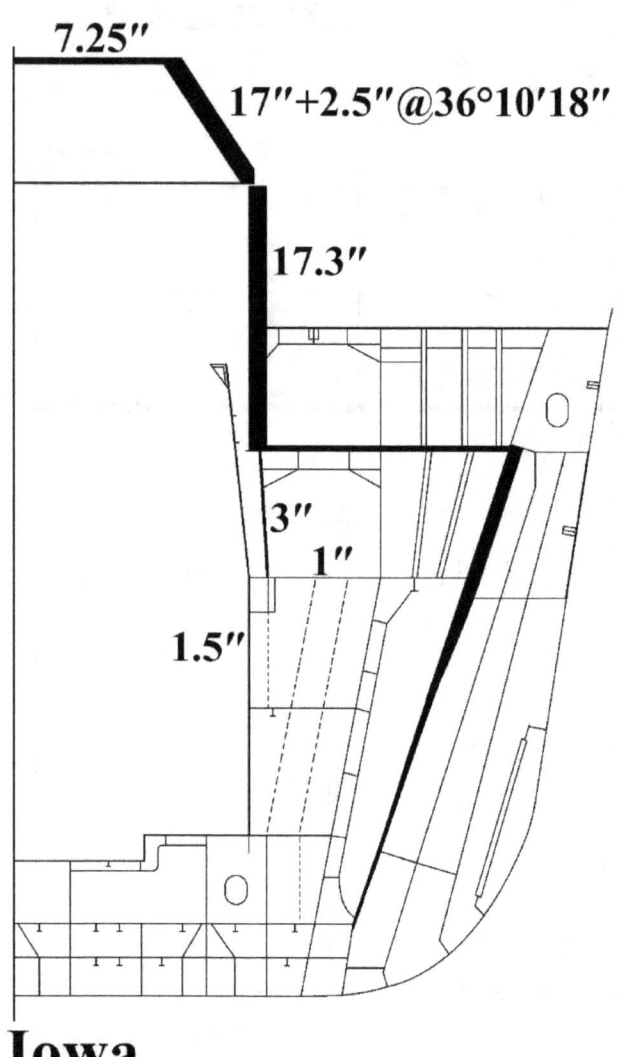

Iowa

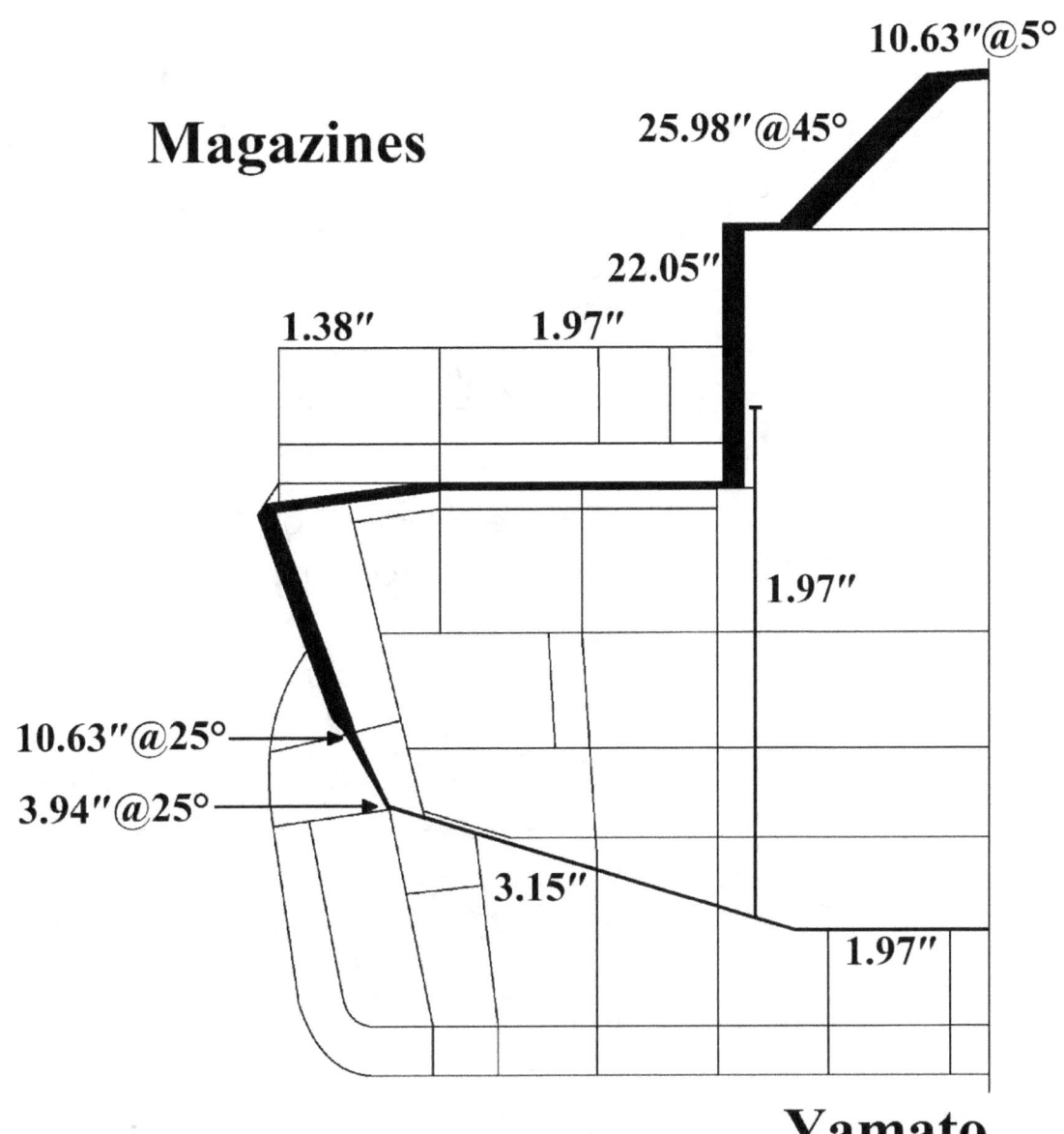

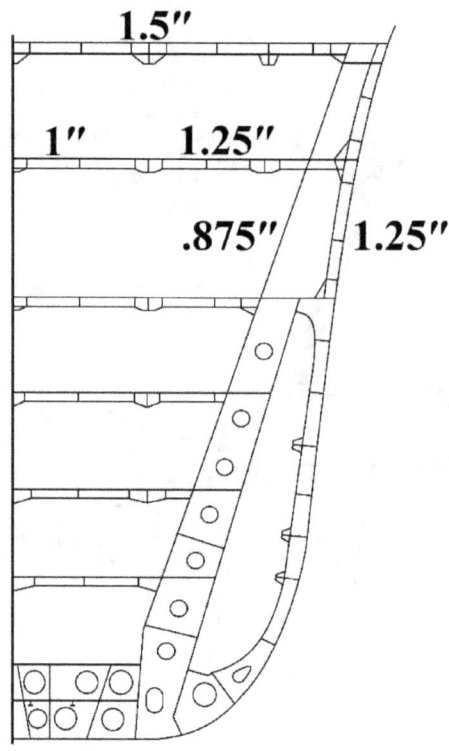

Bow

1.38"   1.97"

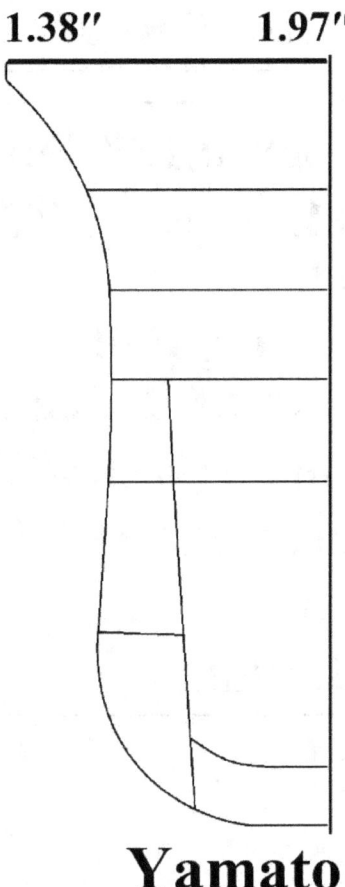

Yamato

Figure 3.3

> A) 6.2" Steering Deck
> B) 1.5" Deck+4.75"+1.25" Deck +.625" Deck+.5" Deck
> C) 7.25" Turret/CT Roof

> I.) 17"+2.5"@36°10'18" Turret Face
> II.) 1.5" Hull+12.1"+.875"@19° Belt +.625" Bhd.
> /13.5"@19° Steering Belt
> III.) 17.3" Barbette/CT Beam
> IV.) 14.8" Barbette (Beam/CL)
> V.) 11.6" Barbette (CL)
> VI.) 11.3" Aft Bhd.
> VII.) 8.5" Fwd. Bhd.

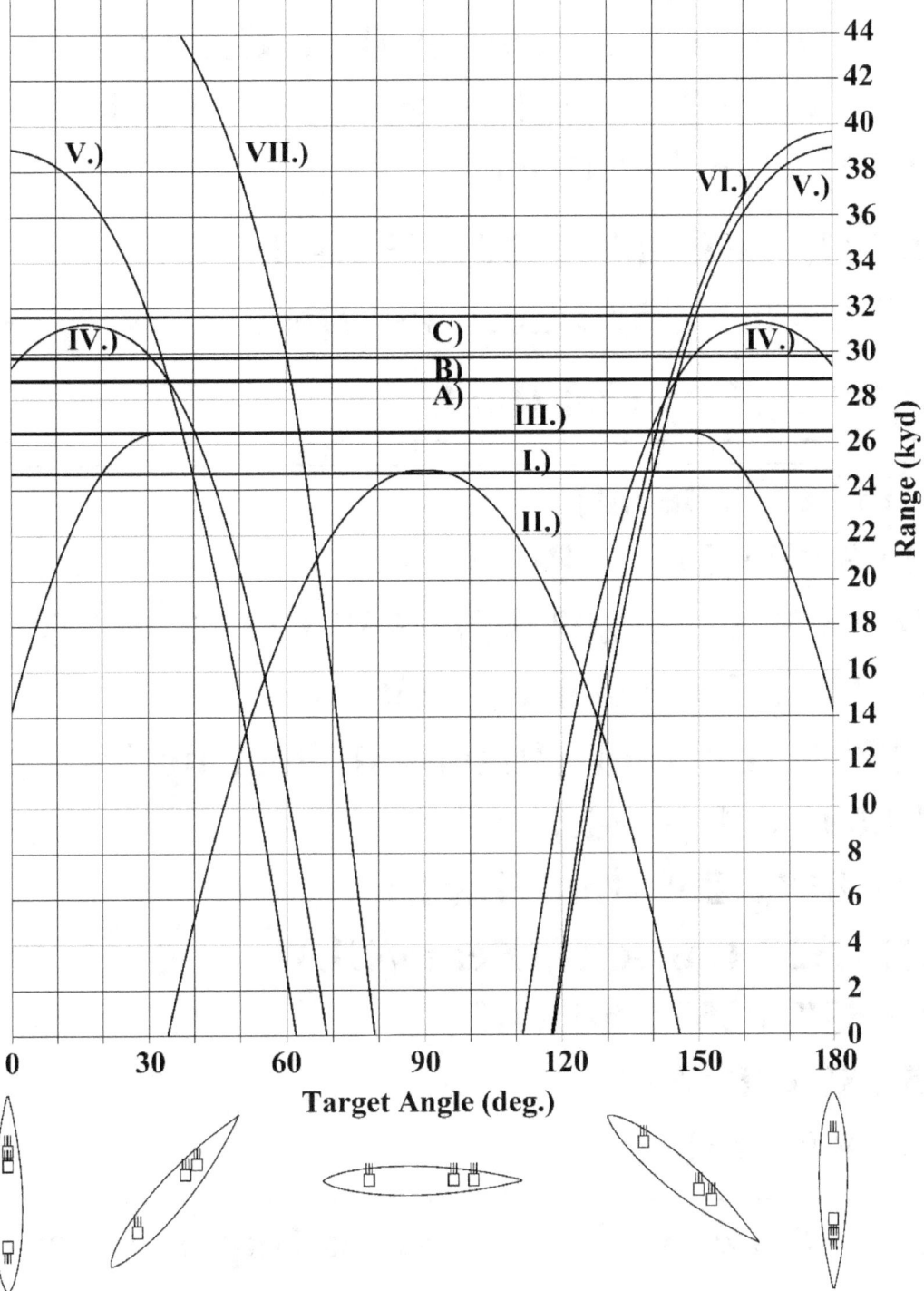

Figure 3.4

A) 7.87" Steering/CT Roof
B) 1.38" Deck + 13.38"@25° Bhd. Crown
C) .47" Deck.+39" Deck +7.87"+.39" Deck +.35" Deck +.35" Deck

  /.98" Hull+ 9.06"+.55"@7° Deck +.35" Deck + .35" Bhd.+.35" Deck

D) 10.63"@5° Turret Roof

I.) 22.05" CT (Beam)
II.) 22.05" Barbette (Beam)
III.) 16.14"+.63"@20° Belt+.55" Bhd.+.35" Bhd.
IV.) 14.17"@25° Steering Belt
V.) 13.78"@25° Bhd. (Diagonal/Steering Fwd.)
VI.) 19.69" CT (Fwd.)
VII.) 11.81"@25° Bhd. (Fwd.)
VIII.) 17.32" Barbette (Beam/CL)
IX.) 10.63"@25° Bhd. (Aft)
X.) 15.75" CT (Aft)
XI.) 14.96" Barbette (CL)

Note: 25.98"@45° Turret Face is Impenetrable.

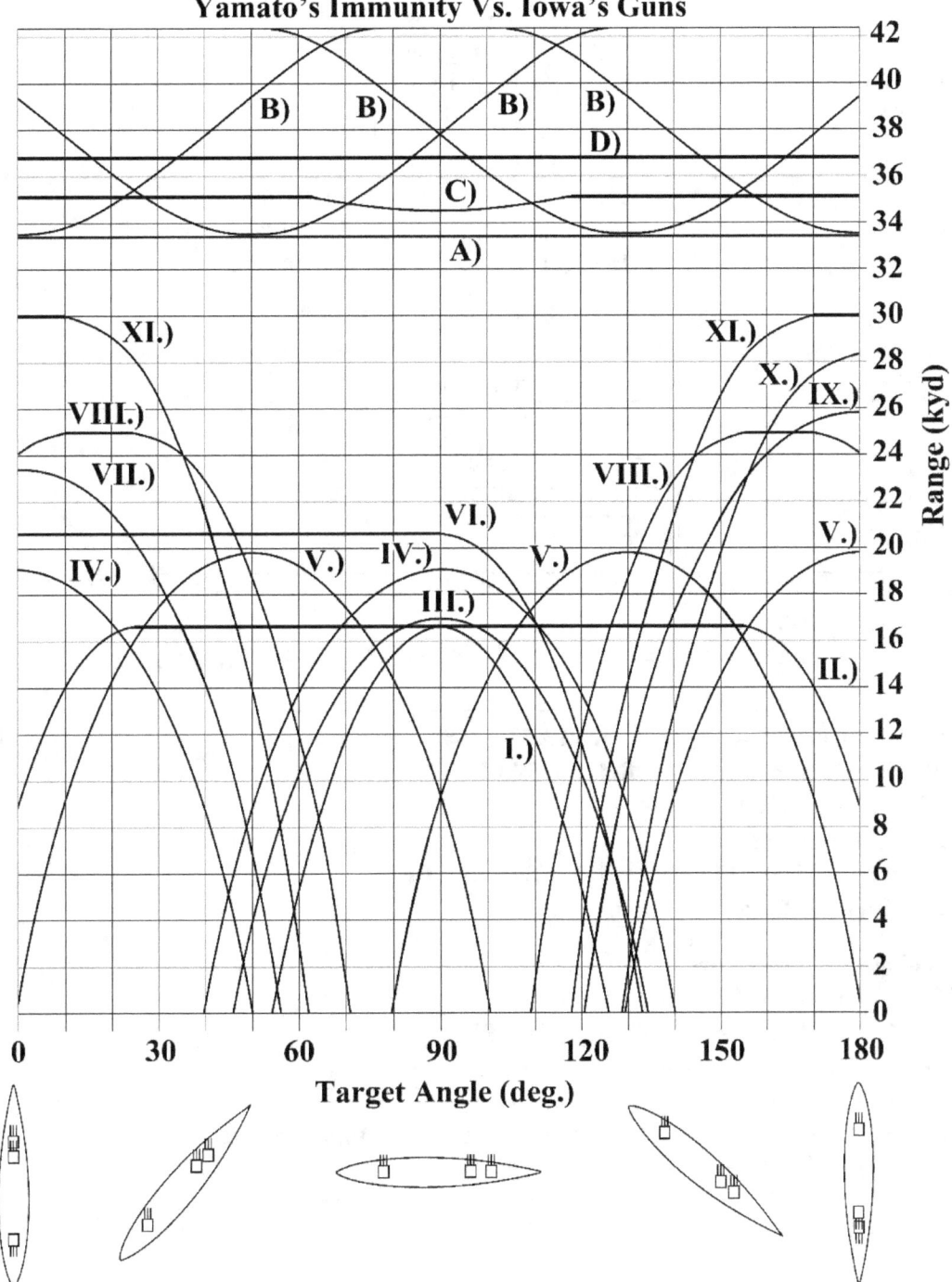

Figure 3.5

Immunity Zones at 90 deg. Target Angle
(Based on Bu. Ord. Sk. 78841)

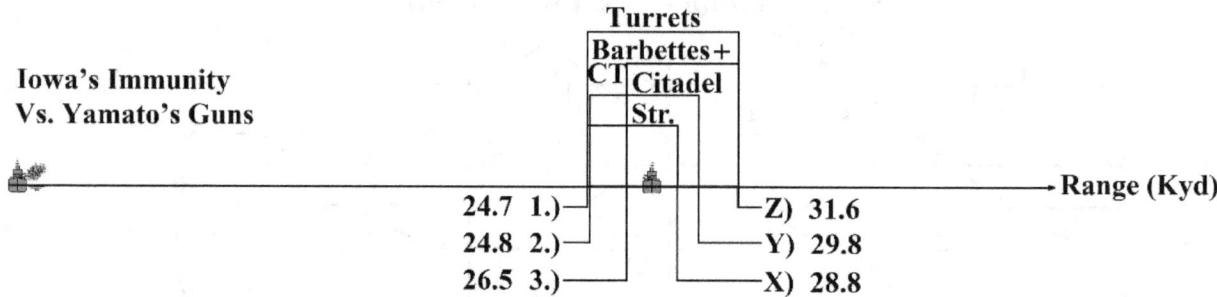

Iowa's Immunity Vs. Yamato's Guns

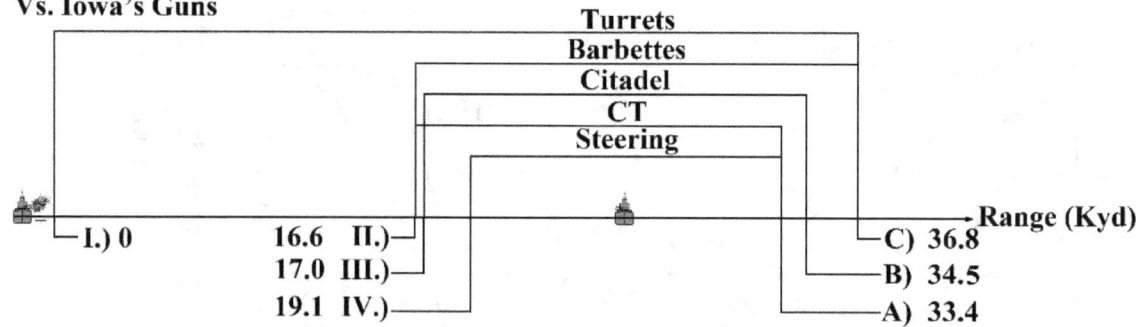

Yamato's Immunity Vs. Iowa's Guns

1.) 17"+2.5"@36°10'18" Turret Face/13.5"@19° Steering Belt
2.) 1.5" Hull+12.1"+.875"@19° Belt+.625" Bhd.
3.) 17.3" Barbette/CT Beam
X) 6.2" Steering Deck
Y) 1.5" Deck+4.75"+1.25" Deck+.625" Deck+.5" Deck
Z) 7.25" Turret/CT Roof

I.) 25.98"@45° Turret Face
II.) 22.05" Barbette/CT Beam
III.) 16.14"+.63"@20° Belt+.55" Bhd.+.35" Bhd.
IV.) 14.17"@25° Steering Belt
A) 7.87" Steering Deck/CT Roof
B) .98" Hull+ 9.06"+.55"@7° Deck +.35" Deck + .35" Bhd.+.35" Deck
C) 10.63"@5° Turret Roof

Figure 3.6

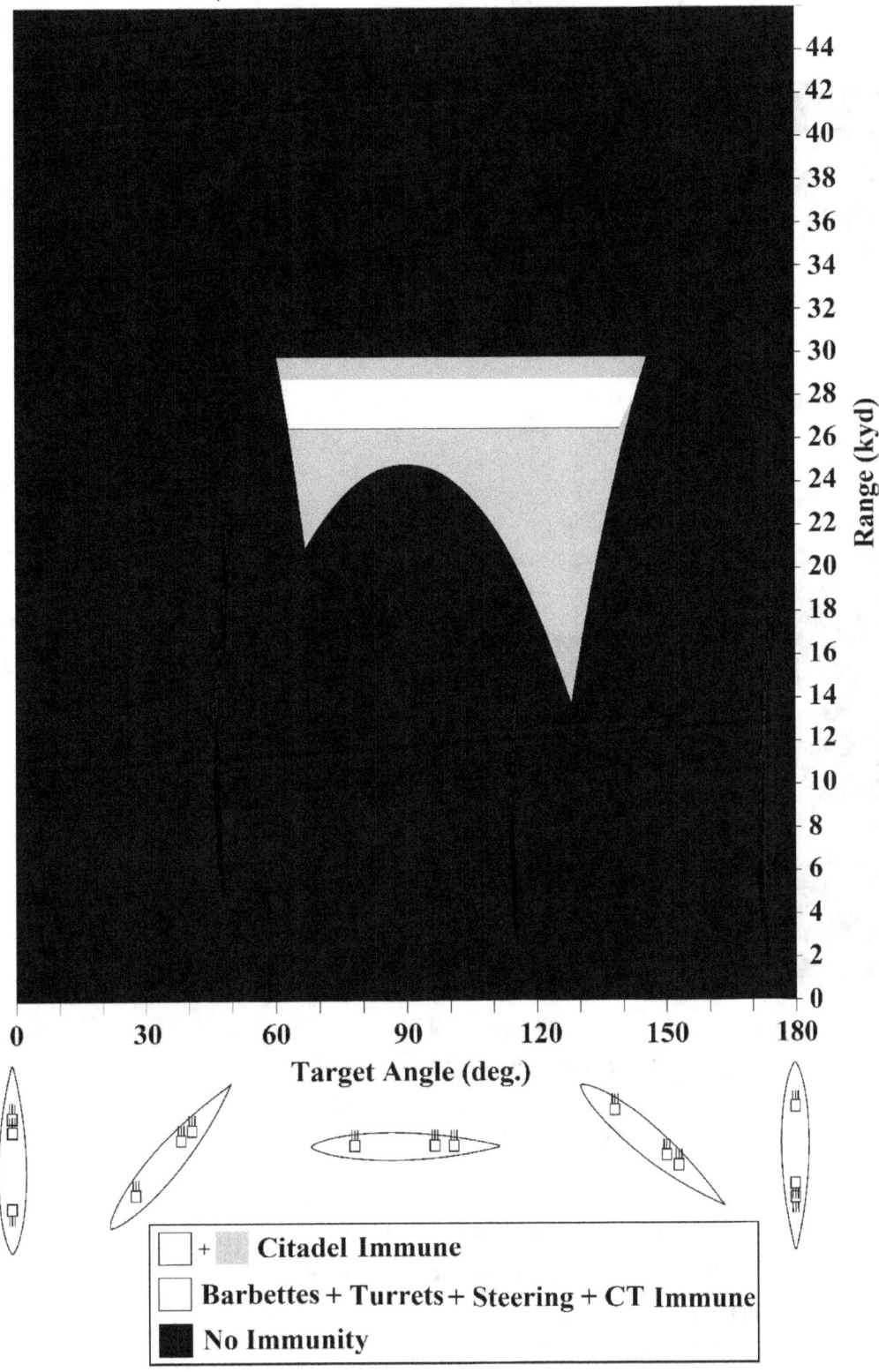

Figure 3.7

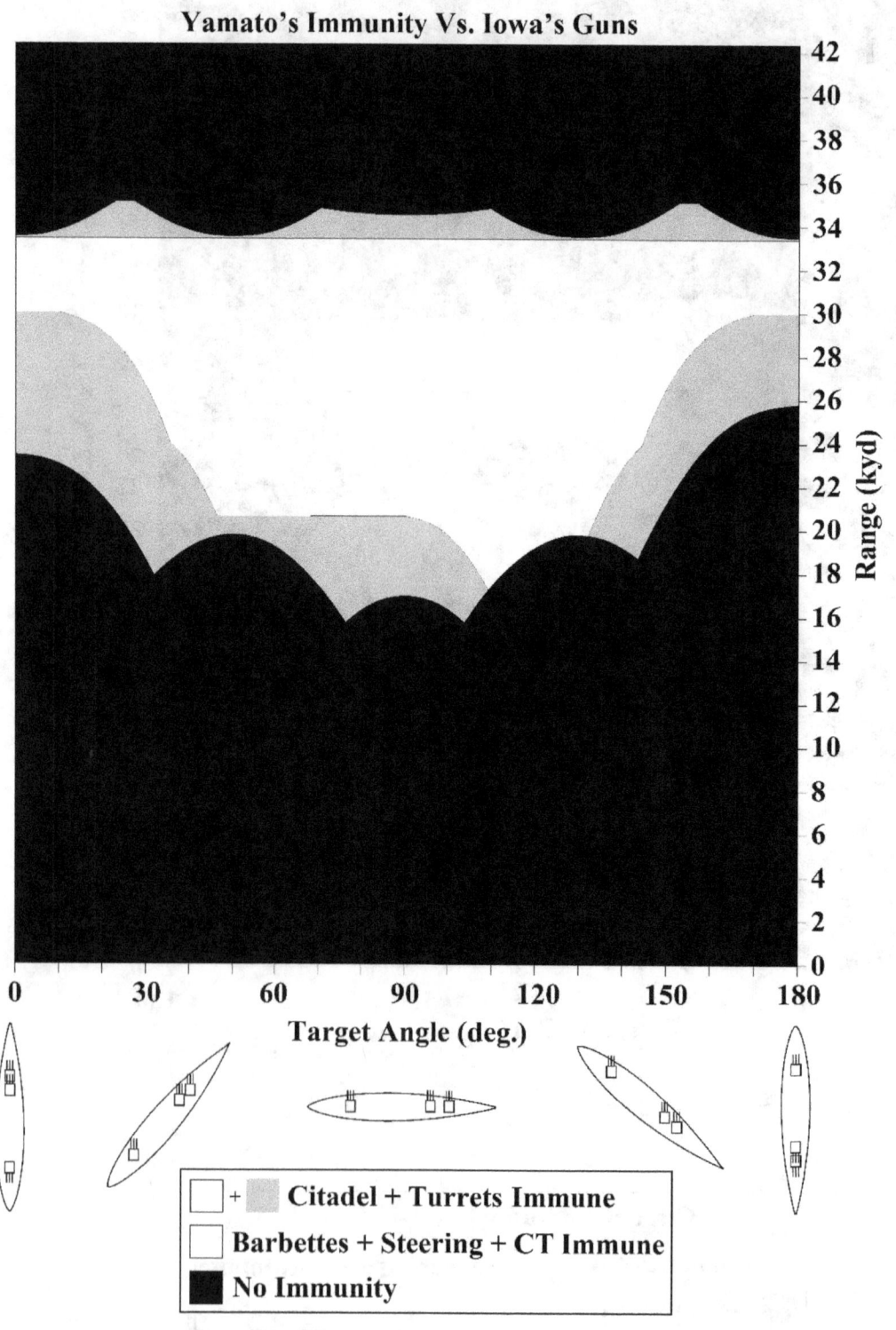

Figure 3.8

Iowa's Immunity
Vs. Yamato's Guns
44
Yamato's Immunity
Vs. Iowa's Guns
42
40
38
36
34
32
30
28
26
24
22
20
18
16
14
12
10
8
6
4
2
0
Range (kyd)
0
30
60
90
120
150
180
Target Angle (deg.)
Iowa)
Yamato)
+ Citadel Immune
Barbettes + Turrets
+ Steering + CT Immune
No Immunity
+ Citadel + Turrets Immune
Barbettes + Steering + CT Immune
No Immunity

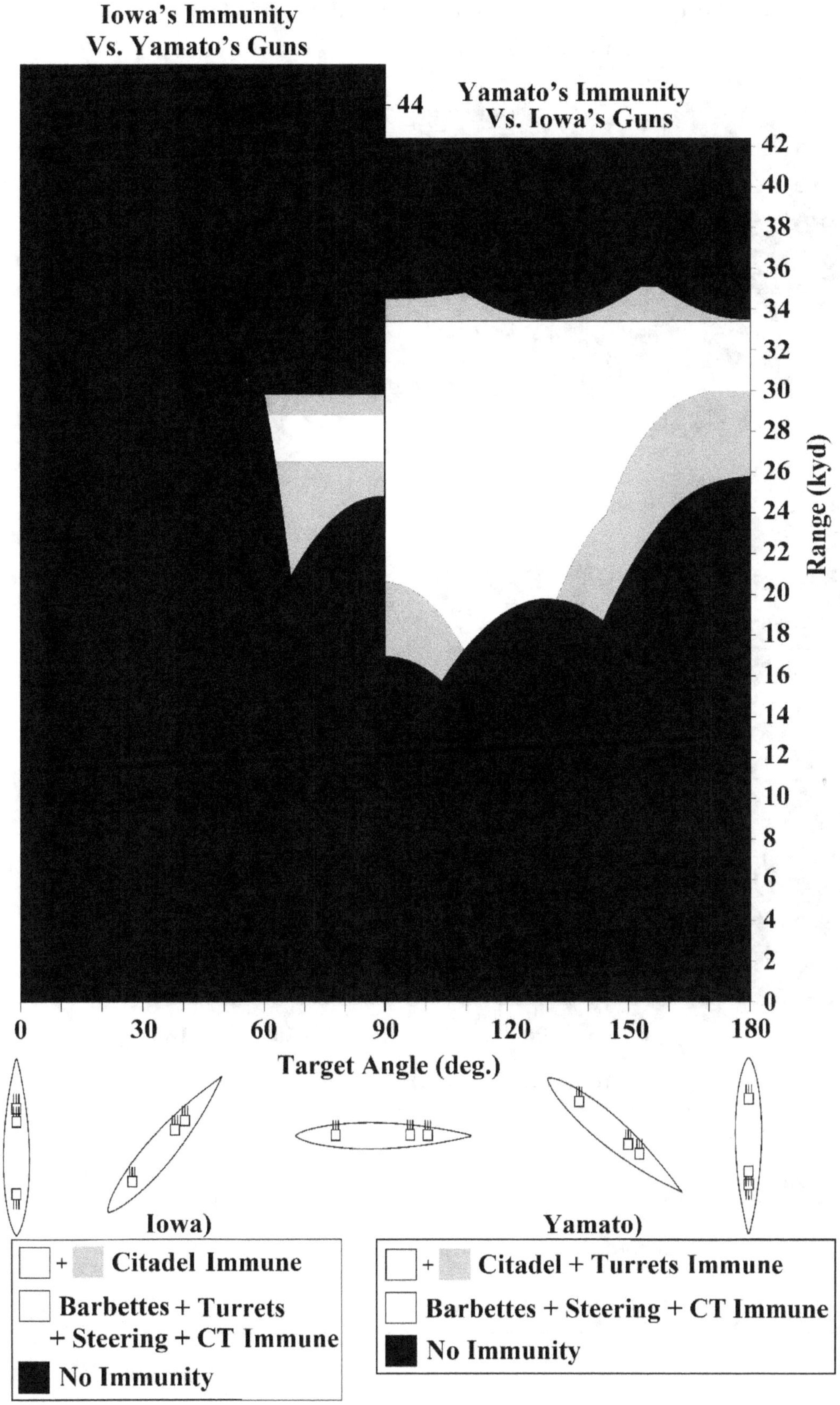

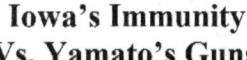

Figure 3.9

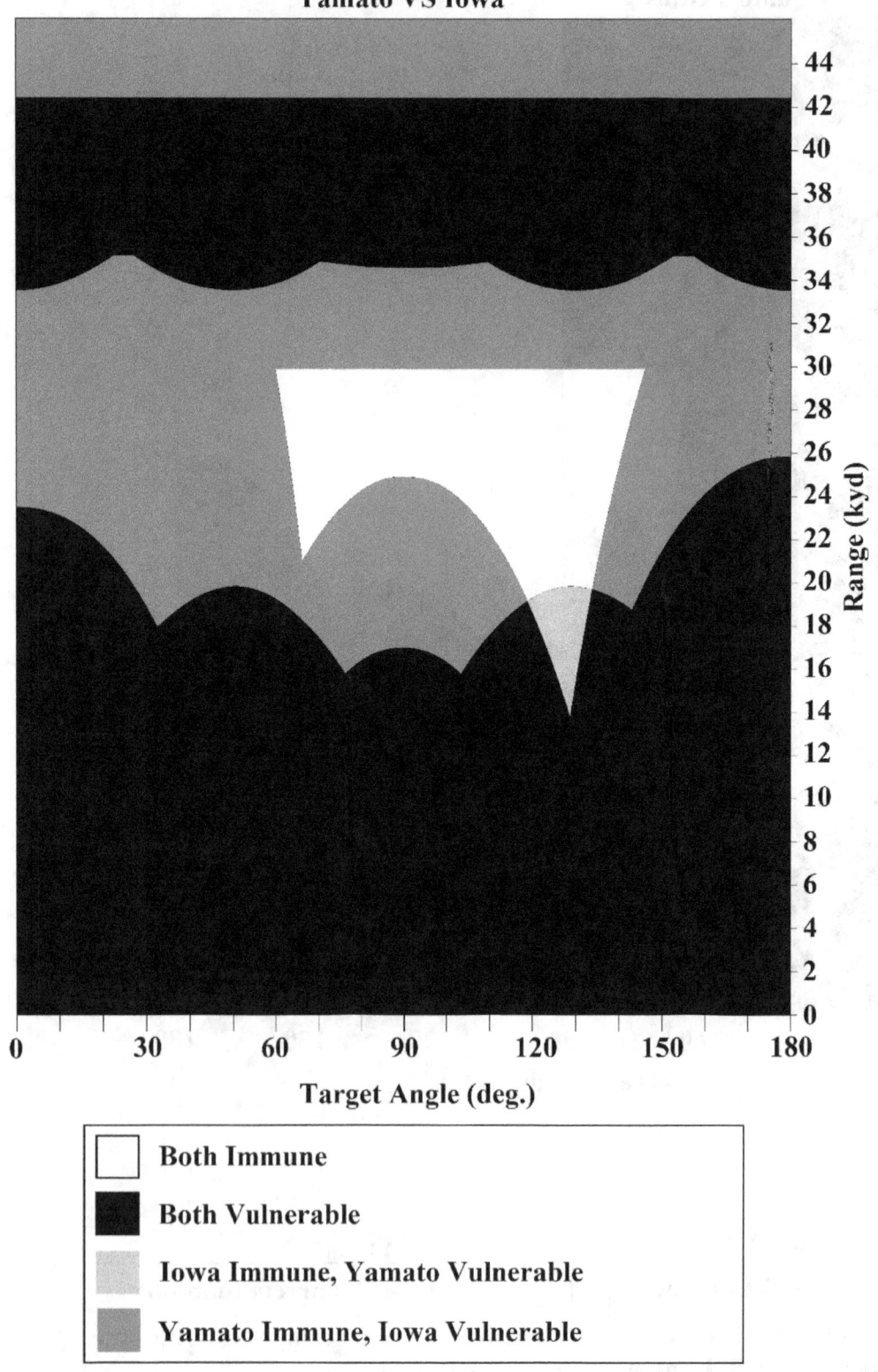

Conclusion

It is of paramount importance that we should, first of all, inquire into the tactical conditions of a possible engagement.

It is apparent that, by virtue of the greater speed of Iowa, it would have been up to the Americans to select fighting range.

It should also be kept in mind that the Americans knew very little about Yamato's technical characteristics, believing that she carried 16" (406 mm) guns and substantially lighter armour than she actually did, although her speed was appraised reasonably well.

Hypothesizing that 16" (406 mm) was the largest caliber afloat, the Americans supposed that the designed immunity of their ship against 16"/2,240 lbs/2,520 fps (406 mm/1,016 kg/768 mps) attack more or less coincided with that of her immunity against the new Japanese ships, while their "super heavy" shells could probably defeat the protection scheme of any ship at optimal fighting range.

Iowa's designed immunity against 16"/2,240 lbs/2,520 fps (406 mm/1,016 kg/768 mps) attack was 18.0-30.0 kyd (16.5-27.4 km), which infers that decisive gun range was estimated at circa 24 kyd (21.9 km). Indeed, during the design period of the preceding ships (South Dakota class), a 1/11 scale-mockup of the ship's side armour was subjected to attack at such obliquity and terminal velocity that corresponds to 14" (356 mm) attack at 23 kyd (21.0 km), hinting that this was the gun range at which the new battleships were expected to engage.

Based on the *USS Massachusetts Gunnery Department Instructions,* the Americans supposed that the new Japanese battleships mounted 16" (406 mm)/45 guns, against which USS Massachusetts was assessed to be immune at 14.0-30.0 kyd (12.8-27.4 km) at 90 deg. target angle. Also, they believed that the new Japanese battleships had a 12" (305 mm) armour belt inclined 20 deg. from the vertical and 6.4" (163 mm) armour deck. Based on the official penetration curves of Iowa, this protection scheme would have given immunity against Iowa's guns at 25.8-29.3 kyd (23.6-26.8 km). Taking into account that the protection scheme of Iowa was almost identical to that of USS Massachusetts, the Americans most likely figured that Iowa could defeat the protection scheme of the new Japanese battleships at 14.0-25.8 kyd (12.8-23.6 km) whilst herself being immune to Japanese attack. However, at 25.8-29.3 kyd (23.6-26.8 km) both ships were immune, while inside 14 kyd (12.8 km), as well as outside 30 kyd (27.4 km), neither. Therefore, the ideal battle range was most likely estimated at about 22 kyd (20.1 km) against the new Japanese ships, despite the fact that the designed immunity of Iowa against domestic guns indicates that optimal gun range was considered to be somewhat greater than this, namely 24 kyd (21.9 km).

Note that American 16" (406 mm) battleships engaged with enemy battleships at such ranges during World War II – USS West Virginia opened fire at Yamashiro at a range of 22.8 kyd (20.8 km); USS Massachusetts engaged Jean Bart at 24.0 kyd (21.9 km). (USS Washington defeated Kirishima at significantly shorter ranges than these, but this engagement hardly went according to preliminary planning; hence it has little representative value.)

Therefore, it is not reasonable to presume that the engagement would have taken place at ranges manifestly shorter or longer than these. For instance, inside 15 kyd (13.7 km), as well as outside 35 kyd (32.0 km), even 14" (356 mm) guns could defeat Iowa's citadel. Hence, under such circumstances, even the old 14" (356 mm) Japanese capital ships could have scored fatal hits against her, even though she could have defeated any of these ships within her designed immunity zone without having to expose herself to anything more than superficial damage.

Furthermore, with the maximum detection range of shell splashes of the Mark 8 radar limited at circa 35 kyd (32.0 km), Iowa could not have fully utilized this asset outside this range, nor could she use her rangefinders – i.e. both sides would have had to rely on air spotting. This, combined with the extremely long time of flight and dispersion of salvos, results in an engagement outcome inevitably characterized by randomness, which is evidently undesirable, especially to the side with the superior ship.

It is also worth pointing out that no battleship gun ever scored a hit in action against a moving target outside 27 kyd (24.7km); and, as we have indicated above, most historic engagements were fought at shorter ranges than this.

Based on the designed immunity of Yamato, the Japanese envisaged a decisive gun range not materially greater than the Americans visualized, namely 27.4 kyd (25.0 km). Granting the fact that Yamato lacked the speed to dictate the tactical conditions of the engagement, it seems unlikely that she would have tried to engage at shorter or appreciably longer ranges than that preferred by the Americans.

To sum up, it seems most probable that decisive gun range would have been somewhere between 22 and 24 kyd (20.1 and 21.9 km) based on the designed immunity of the ships, the fighting range at which similar historical engagements took place, and the preliminary planning of the Americans based on the guesstimated technical characteristics of Yamato as described by the *USS Massachusetts Gunnery Department Instructions*.

However, this makes it exceedingly difficult to avoid the conclusion that this engagement, had it occurred, would have almost certainly had a disastrous outcome for the Americans, given that, in reality, it was Yamato that was completely secured and could defeat the protection scheme of her foe under these circumstances, not Iowa; the Americans grievously underestimated the new Japanese battleships. As illustrated above, Bu. Ord. Sk. 78841 predicts not only that Iowa's period of immunity was such that it could have been easily cancelled out by the inherent motions of the ship and the varying performance of materials, but it was also almost completely outside the fighting range thought to be optimal; not to mention that the official penetration values of the Japanese gun hint that Iowa had no immunity at all. Furthermore, Yamato could unleash 36 % more energy, 28 % more steel, and as much as 116 % more TNT equivalents (49 %, 37 %, and 153 %, respectively, if Yamato's six medium caliber guns are also considered) in total as a function of time as compared to Iowa. The difference between the effects of a given amount of explosive detonating outside a ship as compared to over twice that amount detonating inside of it is, needless to say, considerable.

A number of theories can be introduced that might allow one to escape this conclusion, allowing that none of these are supported by consistent empirical data, nor are overwhelmingly convincing. Still, speculative though they may be, it is worthwhile to list them to make our inquiry complete:

- <u>If we assume that Iowa's hull plating knocks off the cap of incoming AP type shells before they make contact with her heavy side armour, then the ship's period of immunity might increase.</u>
  While this can be true, even if we grant that AP shells are decapped and their limit velocity increases as suggested by ADM 281/31 and 281/37, the ship's period of immunity still remains too small to be truly reliable. Equally problematic is that transverse bulkheads, turret faceplates, barbettes, the sides of the conning position, and that of the steering gear were not shielded by thickened hull plating. Therefore, they could still be perforated at the ranges specified above. Furthermore, the thickened portion of the hull plating terminated at the waterline. Consequently, projectiles diving no more than a few feet could have made contact with the lower belt of the ship without even having to defeat the thickened portion of the hull. Now, it is not too difficult to see why this would have been particularly problematic against the Japanese shells that were specifically designed for attacking the underwater hull of enemy ships. Also, it is highly doubtful whether a 1.5" (38 mm/0.08 cal.) vertical plate can decap an 18.07" (459 mm/1.0

cal.) AP projectile in the first place. On top of that, ADM 281/31 implies that decapping plates are not effective if projectile diameter materially overmatches armour thickness:

"Against super calibre attack a spaced assembly is inferior as regards perforation velocity to a solid plate of the same overall poundage."

Finally, based on the official penetration values of Yamato, Iowa had no immunity even if projectiles are assumed to be decapped.

- <u>The inner limit of Iowa's immunity would have been more favorable if she engaged at, say, 60 deg. target angle.</u>

  Despite being a more persuasive argument than the preceding one, most objections raised against it still hold true. The immunity of turrets, turret barbettes, and that of the conning position cannot be enhanced by inclining the ship; and the official penetration values of Yamato indicate no immunity at 60 deg. target angle either, with the armour deck being vulnerable against the completely flat, 670 BHN projectiles outside circa 22 kyd (20.1 km) and the belt up to 23.5 kyd (21.5 km) at 60 deg. Now, the thickness of the transverse bulkheads and the centerline arcs of barbettes was such that it seems unlikely that the Americans envisaged engaging at less than 60 deg. target angle. In fact, even Bu. Ord. Sk. 78841 implies that at 60 deg. target angle, the lower strake of the forward transverse bulkhead, if hit through the bow, was already vulnerable, as was the vertical armour of the turrets, turret barbettes, and the conning position at the ranges the engagement most likely would have taken place.

  Needless to say, Yamato might not have engaged at 90 deg. target angle either, further increasing her already immense period of immunity.

- <u>If we presuppose that Japanese heavy armour was of inferior quality, we might surmise that the immunity zone of the Japanese ship was smaller than indicated by Bu. Ord. Sk. 78841.</u>

  While it is correct that NPG report No. 5-47 suggests that Japanese armour had inferior effective resistance as compared to the predictions of Bu. Ord. Sk. 78841, it should be noted that, when quantified, the difference between the performance of Japanese armour and the predictions of this formula is quite negligible. Based on the 13 impacts listed in this report, the limit velocity of the Japanese armour plates, on average, was only 1.8 % below the predictions of Bu. Ord. Sk. 78841. Even if we postulate that the mean performance of the armour plates fired at in the U.S. are identical to those carried by Yamato, the ship's immunity would shrink very little as compared to what is given above.

  It should also be pointed out that the British, based on 6 impacts, found Japanese heavy armour to be of exceptional quality, and superior to both German types and those of their own in terms of effective resistance. Hence, if we base our calculations on the British tests results, then Yamato's immunity would increase, not decrease.

  Furthermore, based on 21 impacts described by NPG report No. 5-47 and A.T.C. 22ND July 1948, the mean performance of U.S. armour was also 1.3 % below the predictions of Bu. Ord. Sk. 78841.

  Arguments alluding to the performance of materials are always highly speculative. This is well-illustrated by the fact that the Americans and the British draw contradictory conclusions based on post-war firing trials regarding the performance of German cemented, non-cemented, as well as Japanese heavy, armour. This is largely attributable to the facts that the performance of armour plates and projectiles is dependent on test conditions, and that the mechanical properties of nominally identical materials were, in reality, scarcely identical. Although all manufacturers strove to produce high quality armour, inevitably some plates produced were lower quality.

  The results of ballistic trials at our disposal suggest that there was no decisive difference between the average performance of armour plates manufactured by the major navies of the period; and the mean performance of the plates produced by any of them, given that the sample is large enough, scarcely deviates more than a few percent from the predictions of Bu. Ord. Sk. 78841.

Hence, it seems justified to consider the predictions of this formula at least good approximations; but it is understood that armour penetration is virtually impossible to model with perfect accuracy, owing to the complexity of the phenomenon and the wildcards addressed above.

- <u>If we surmise that the later mods. of the U.S. AP projectiles had superior penetrative capability, then it is possible that Yamato's immunity was smaller.</u>

  Based on NPG report No. 3-47, the Americans attained enhanced penetrative capability with "super hard" (680 BHN) projectiles. However, according to the report, only medium, no major, caliber projectiles were manufactured with such maximum hardness. The report reveals that the detailed hardness distribution of the AP Mark 8 mod. 6 (introduced in 1944) was roughly the same as those of previous major caliber projectiles, i.e. a "sheath hardened" shell body and cap with a maximum hardness of "only" 555 BHN.

  Perhaps the most noteworthy difference between the earlier and later mods. of the AP Mark 8 was that the newer shells were blunter, as was their armour-piercing cap. The armour-piercing cap of the mod. 6 projectile shown by this document, for example, was a cone with a half nose-angle of circa 55 deg. NPG report No. 2-43 and the document titled *Effect of Projectile Nose Shape on Ballistic Limit Velocity, Residual Velocity, and Ricochet Obliquity* describe the ballistic performance of shells with similar nose shape. The former document highlights the performance of conical-nosed projectiles with a half nose angle of 68 deg. and 51 deg. as compared to more tapering, ogival projectiles, against homogeneous armour plates ranging in thickness from 1/4 to 4/3 cal. at obliquities ranging from 0 to 45 deg. The test results evince that the conical shells were not superior to the ogival ones under these conditions. However, the latter document revealed that projectiles with 60 deg. half-nose angle had superior performance in cases of low T/D ratio as compared to projectiles with a half-nose angle of 30 deg. (The reasons for this are addressed in Chapter 1 and Appendix E.) This hints at the possibility that the newer shells had superior performance against horizontal armour. While this might justify the assumption that the outer limit of Yamato's immunity was somewhat smaller, it should be kept in mind that the designed immunity of Yamato was up to 32.8 kyd (30 km) against the harder, larger, and completely flat Japanese projectiles. Based on the same source, completely flat projectiles are superior to both the 30 deg., half-nose angled conical shells, as well as the 60 deg. ones, under the test conditions specified above. Therefore, it seems unlikely that Yamato's outer limit of immunity was less than this against the newer U.S. projectiles. Though less than the circa 35 kyd (32.0 km) outer limit Bu. Ord. Sk. 78841 alludes to, this distance is still well beyond the range the engagement was likely to occur.

Summary of speculations alluding to terminal ballistics.

To sum up all the preceding arguments, we might set up a best-case and a worst-case scenario:

Table C.1

Summary of Speculations Alluding to Terminal Ballistics		
Designation	Best-Case Scenario (for U.S.)	Worst-Case Scenario (for U.S.)
Hull Decaps Projectiles	Yes	No
Horizontal Angle (Iowa)	<90< deg.	90 deg.
Horizontal Angle (Yamato)	90 deg.	<90< deg.
Penetrative Capacity (Iowa)	Deck Penetration Capacity Increased	As Described by Bu. Ord. Sk 78841
Penetrative Capacity (Yamato)	As Described by Bu. Ord. Sk 78841	As Indicated by Official Data
Effective Resistance of Armour (Iowa)	As Described by Bu. Ord. Sk 78841	As Indicated by NPG 5-47+ATC 22nd July 1948
Effective Resistance of Armour (Yamato)	As Indicated by NPG 5-47	As Indicated by ATC 22nd July 1948

Now, in the best-case scenario for the Americans, Yamato is still immune at circa 18-32 kyd (16.5-29.3 km), i.e. she is immune at the optimal fighting range, but she can penetrate Iowa's turret face plates, turret barbettes and conning tower. Allowing that at 60 deg. target angle Iowa's main belt would have been immune against decapped projectiles, the lower strake of her forward transverse bulkheads was not.

In the worst-case scenario, Yamato is almost completely immune up to 36 kyd (32.9 km) occasioned by the high probability of projectile failure upon contact with her heavy side armour at high obliquity, although her vertical plates could have been defeated at very short ranges if hit at right angles. She could pierce Iowa's armour at all ranges.

A scenario somewhere between these two seems most probable. Indeed, our calculations correspond to this assumption.

- Iowa's fire-control system might have been more effective.

Of the most important factors governing the outcome of a gunnery duel, this is by far the most difficult to evaluate in the absence of empirical data; hence it is also the one that leaves the greatest ground for speculation.

The assumption that the Mark 38 system was superior to the Type 98 system cannot be backed by empirical data. Based on USNTMJ OT O-31 *(Japanese Surface and General Fire Control)*, it was within the limits of the Type 98 system to deal with an Iowa class battleship sailing at maximum speed up to extremely long ranges in severe weather conditions whilst the firing ship herself was sailing at maximum speed. The same most likely holds true for the Mark 38 system as well.

Based on the same source, and USNTMJ OT O-29 *(Japanese Fire Control)*, the Americans considered their stable verticals and fire-control switchboards more sophisticated than those of Japanese manufacture, albeit the Japanese devices described in these documents were those of the older, not Yamato class, battleships. Unfortunately, most documents dealing with the equipment of the newest ships were destroyed.

The operation of the U.S. system, however, was definitely made smoother by RPC – something even the newest Japanese battleships lacked. Admittedly, no British battleship prior to HMS Vanguard had RPC either; yet, using analogous "follow the pointer" systems as the Japanese did, they proved to be no less capable of delivering accurate fire than their U.S. contemporaries.

Arguably more important, at least under certain conditions, was the edge the U.S. had in radar technology compared to Japan. Notwithstanding that the more advanced fire-control radars of the U.S. ship were, to some extent, countered by the superior optics of Yamato in good visibility, in case the engagement would have taken place at night, the Americans would have had an edge in target information acquisition. This being said, Yamato was not completely blind at night either, carrying two sets each of three different types of radars, including the combined surface search/fire-control set delineated above, as well as long-range starshells, searchlights, infrared (yamakawa) light sources, etc. Still, it is not groundless to suppose that fire-control efficiency at night on the American side would have surpassed that of their adversary by virtue of the greater accuracy of the U.S. fire-control radar. However, this does not change the fact that any damage sustained on the Japanese side would have been superficial at the range the engagement most likely would have taken place, whilst hits scored by Yamato could easily have been fatal, even if fewer in number.

Figure C.1

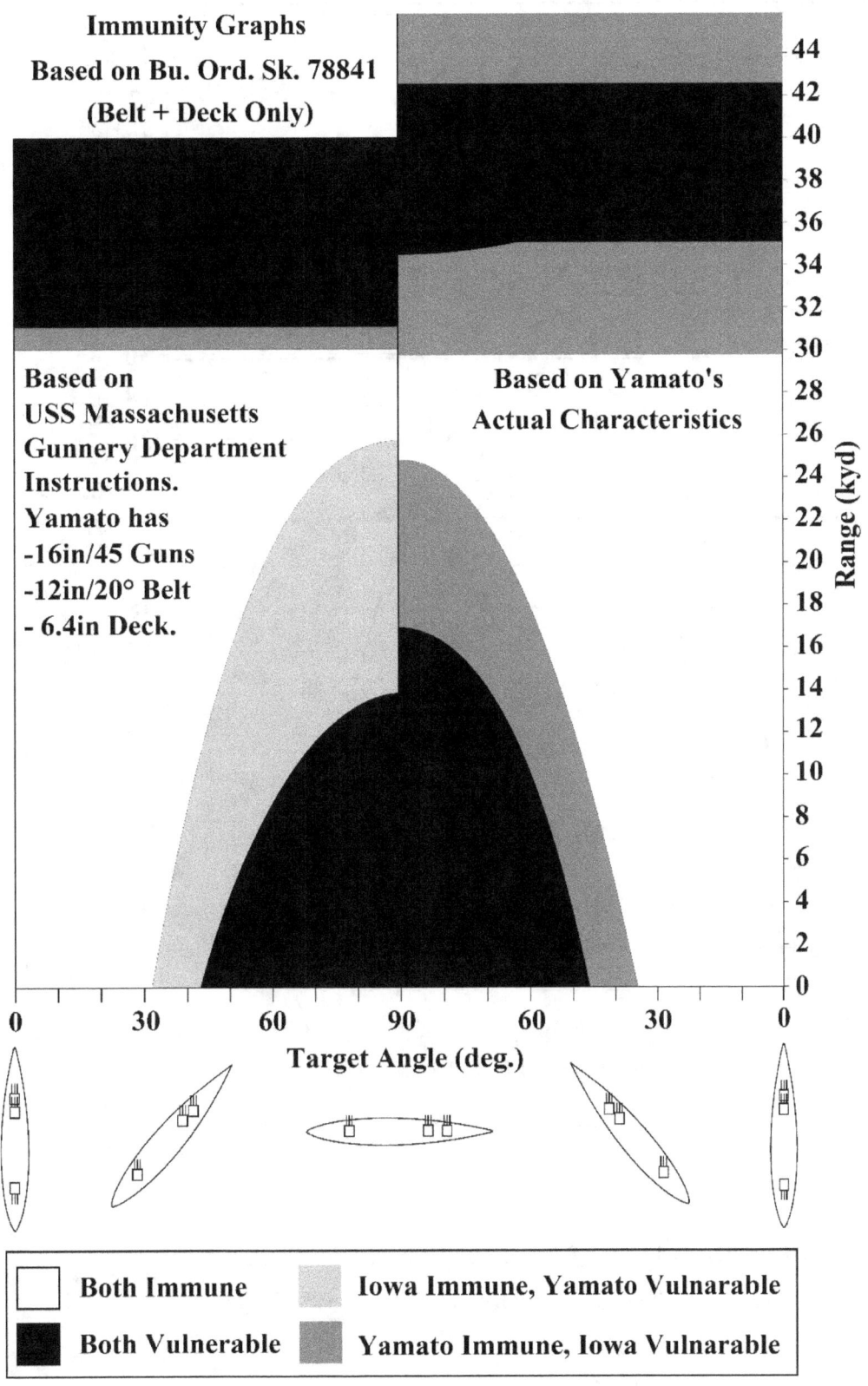

Figure C.2

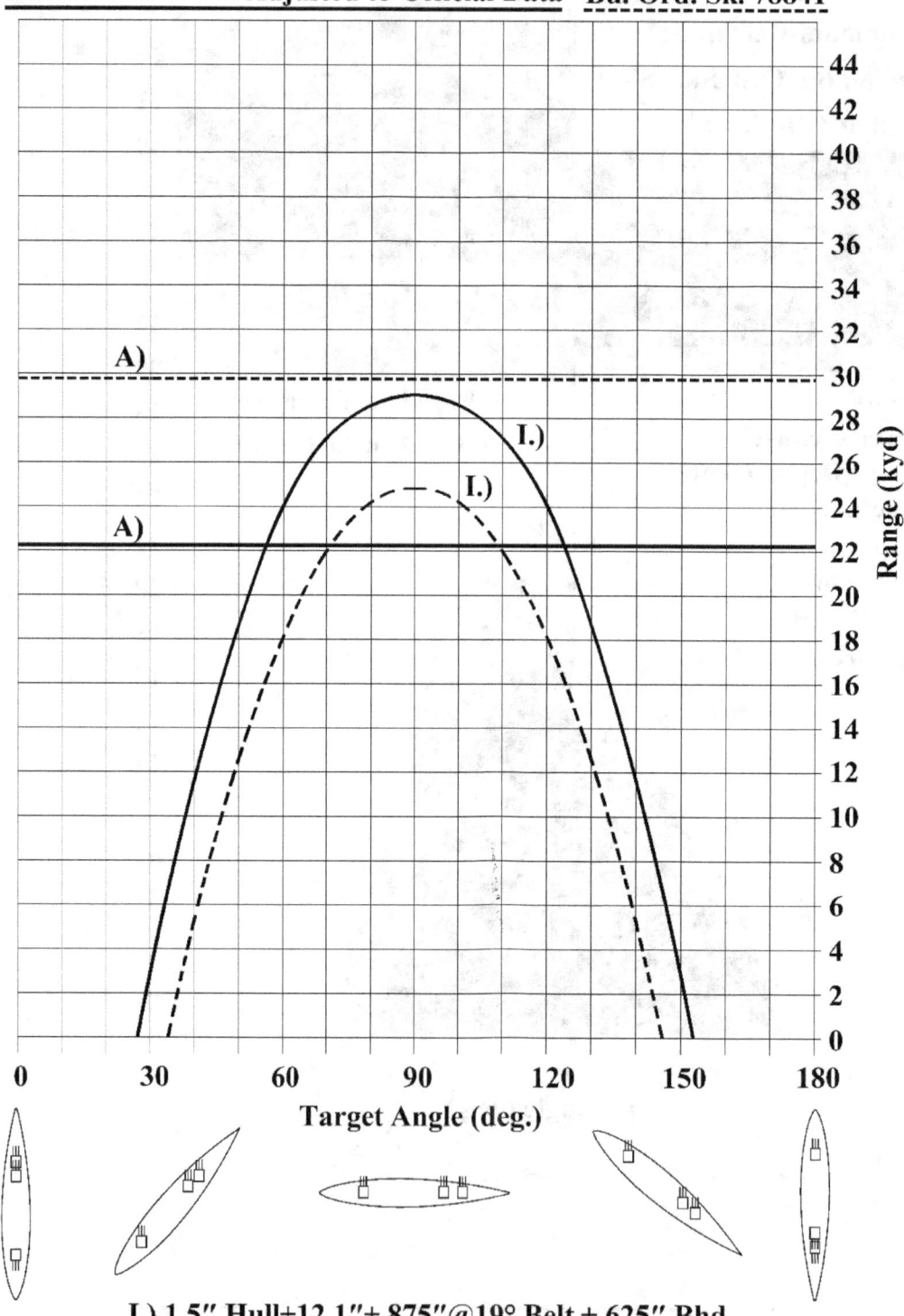

I.) 1.5" Hull+12.1"+.875"@19° Belt +.625" Bhd.
A) 1.5" Deck+4.75"+1.25" Deck+.625" Deck+.5" Deck

Figure C.3

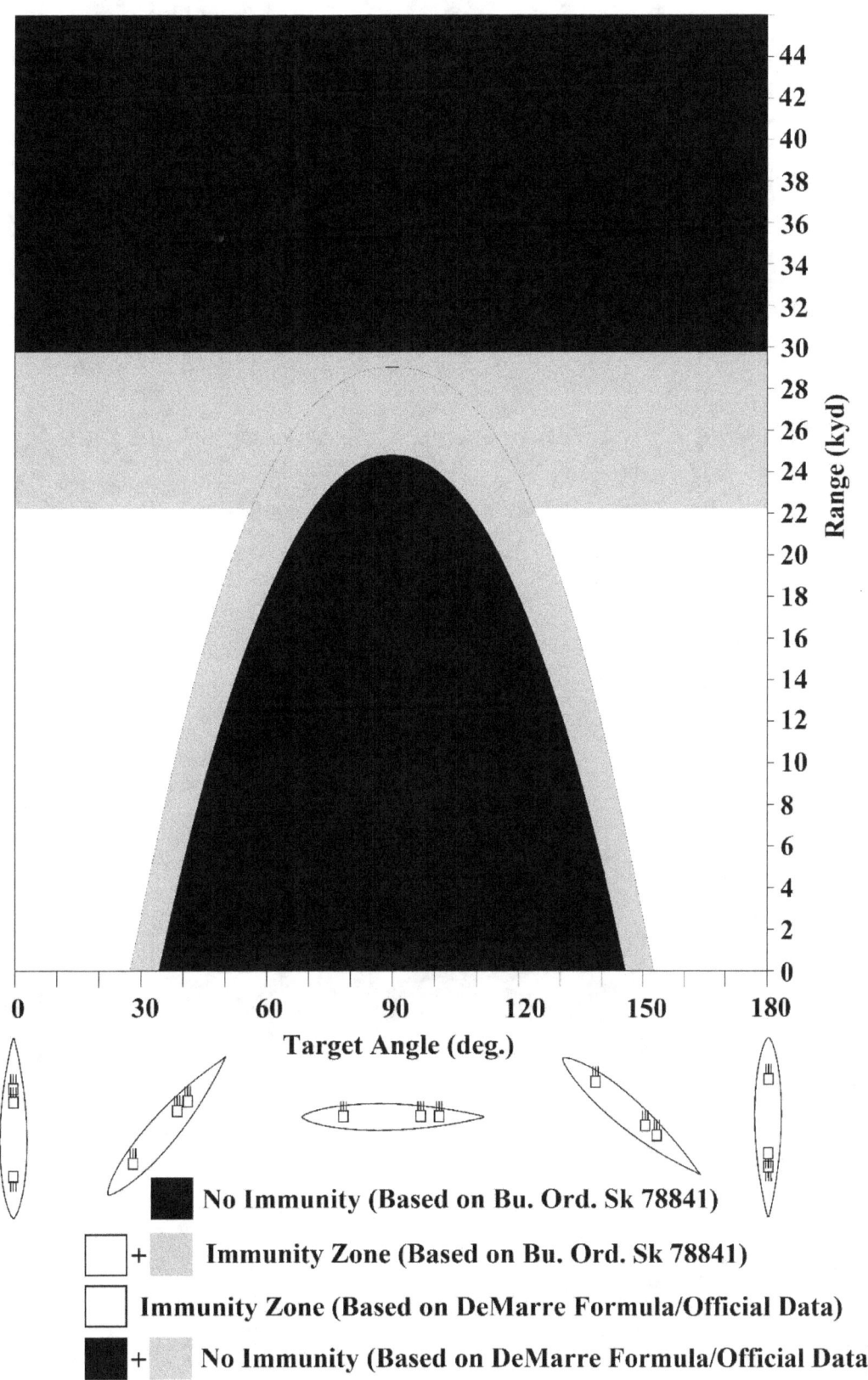

Appendix A – General Characteristics (WWII)

Table AA.1

Designation	Iowa	Yamato
Cost – USD	100 million*	70 million**
Laid Down	27 June 1940	4 November 1937
Launched	27 August 1942	8 August 1940
Commissioned	22 February 1943	16 December 1941
Displacement (Full Load) – tons (m.t.)	57,540 (58,463)	69,988-71,658 (71,111-72,808)
Length (Overall) – ft (m)	887.2 (270.4)	862.9 (263.0)
Length (Waterline) – ft (m)	859.5 (262.0)	839.9 (256.0)
Beam – ft (m)	108.2 (33.0)	127.6 (38.9)
Draught (Full Load) – ft (m)	37.8 (11.5)	35.8 (10.9)
Boilers	8 x Babcock & Wilcox	12 x Kampon
Turbines	4 x General Electric	4 x Steam Turbines
Machinery Output (Designed) – shp	212,000	150,000
Machinery Output (Overload) – shp	254,000	165,000
Speed – knots	33.0	27.46
Fuel Capacity – tons (m.t.)	8,624 (8,762)	6,201 (6,300)
Electric Plant Output – kW	10,000	4,800
Range – Nmi	15,900 at 17 knots	7,200 at 16 knots
Armament	9 x 16" (406 mm)/50 Mark 7 (3x3)	9 x 18.11" (460 mm)/45 caliber guns (3x3)
	20 x 5" (127 mm)/38 Mark 12 (10x2)	12-6 x 6.10" (155 mm)/60 caliber guns (4-2x3)
	60-78 x 1.57" (40 mm)/56 Bofors (15-19 x 4)	12-24 x 5.00" (127 mm)/40 caliber guns (6-12x2)
	52-68 x 0.79" (20mm)/70 Oerlikon (52-60x1+0-16x2)	24-156 x 0.98" (25 mm)/60 caliber guns (8-50x3+0-6x1)
Armour – tons (m.t.)	18,214 (18,506)	22,533 (22,895)
Designed Immunity	18.0-30.0 kdy Vs. 16"/2,240 lbs/2,520 fps Gun (16.5-27.4 km Vs. 406mm/1,016 kg/768 mps)	21.9-32.8 kyd Vs. 18.11"/3,219 lbs/2,559 fps Gun (20-30 km Vs. 460 mm/1,460 kg/780 mps)
Aircraft	3 x Vought OS2U Kingfisher or Curtiss SC Seahawk	7 x Mitsubishi F1M2 Type 0
Complement	Up to 2,788	Up to 3,332
Radar	FC Marks 3 or 27, 4 or 12/22, 8 or 13, 29 SG, SK, SQ, SR, SU	Types 13, 21, 22

*=100 million USD (1943) =1.53 billion USD (2021).
**=250 million JPY (1941) =70 million USD (1941) =1.26 billion USD (2021).

Appendix B – Calculating Penetration Values

The mathematical formula we apply to calculate immunity zones is based on the following assumptions:
1) We assume that total kinetic energy required to defeat spaced assemblies equals the sum of the individual energies needed to defeat each individual element.
2) We assume that in cases of laminated armour effective thickness equals the thickness of the thickest plate plus 70% of the thickness of the thinner plates.
3) Based on empirical data extracted from ADM 281/31 and ADM 281/37, we assume that the ballistic limit velocity of armour plates increases in proportion to the weight of armour piercing caps in cases of decapped projectiles. We assume that projectiles are decapped by armour plates thicker than 0.15 cal.

The relevant passages of ADM 281/37, which specifically deals with major caliber projectiles, is cited below.

1. INTRODUCTION
...It was decided to investigate the performance of decapped shell against various types of deck armour. For this purpose, 200 lbs./sq. ft. and 240 lbs./sq. ft. plates were ordered.

2. OBJECTIVE OF TRIALS
This was: -

(a) To compare the ballistic performance of non-cemented deck armour under attack by 15" A.P.C. at 60° and 65° with the shell in the capped and decapped conditions...

6. ANALYSIS OF RESULTS
...Penetration velocities for the 200 lbs/sq. ft. plates are shown in Table 1.

	TABLE 1		
Angle of Attack	Capped shell Wt. 1938 lbs.	Decapped Shell	
		Wt. 1712 lbs.	Wt. 1938 lbs.
60°	980 ft/sec.	1160 ft/sec.	1090 ft/sec.
65°	1030 ft/sec.	1285 ft/sec.	1210 ft/sec.

The figures for the capped shell are the mean of the results obtained for Investigation No. 8. The decapped shell weighs 1712 lbs. and the figures given in this column are the mean of the results obtained in the present Investigation. The figures for the decapped shell weighing 1938 lbs. are obtained from the results for the 1712 lbs. shell by assuming that the energy required by the shell to achieve penetration is the same i.e. mv^2=constant.

There is a marked difference between the performance of the capped and decapped shells and this still exists for capped and decapped shells of the same weight. An explanation of this may be that the shoulder of the cap succeeds in biting into the plate, even at such oblique angles as 60° and 65°, and, even though the cap may break, the shell will have been given an angular velocity tending to bring it towards the plate normal. The decapped shell on the other hand strikes on the comparatively flat shell shoulder and it is not until the shoulder has dented the plate that the shell is given a turning moment. Hence the decapped shell has a greater tendency to ricochet.

Cracking occurred in the 200 lbs/sq. ft. plates when attacked by the decapped shell, illustrating the fact that cracking is not caused entirely by the cap of a shell.

Table 2 gives similar information for the 240 lbs/sq. ft. N.C. plates: -

TABLE 2			
Angle of Attack	Capped shell Wt. 1938 lbs.	Decapped Shell	
		Wt. 1712 lbs.	Wt. 1938 lbs.
60°	1225 ft/sec.	>1400 ft/sec.	>1320 ft/sec.

The figure for the capped shell is the standard velocity for the given plate thickness and attack. The trend is the same as for the 200 lbs/sq. ft. plates.

The results for the C. and F.H. plates are very similar, (except for plate 4191 which broke up) and a penetration limit velocity of 1260 ft/sec. is a good representative value. This is well below the value of >1400 ft./sec. for the N.C. plates...

7. CONCLUSION

...(3) The penetration limit velocity for the de-capped 15" A.P.C. shell is as much as 200 ft./sec. above that for the capped shell when attacking 200 and 240 lbs./sq. ft. plates at 65° and 60°...

Summary of results obtained with 15" A.P.C. decapped shell

PLATE NO.	TYPE	ANGLE OF ATTACK	PENETRATION VEL.	TYPE OF CRACKING
8432	200 lbs. N.C.	65°	<1226	Heavy
8433	200 lbs. N.C.	60°	1160	Heavy
8434	240 lbs. N.C.	60°	>1437	Slight
8438	240 lbs. N.C.	60°	1370	Slight
4187	200 lbs. N.C.	65°	>1343	Slight
4186.A	200 lbs. N.C.	60°	1155	Moderate
4188	240 lbs. F.H.	60°	1220	Slight
4189	240 lbs. F.H.	60°	<1276	Moderate
4190	240 lbs. C.	60°	1280	Slight
4191	240 lbs. C.	60°	-	Heavy (Broke up)

ADM 281/37 indicates that limit velocity increases – on average – in proportion to the weight of the cap:

Table AB.1

Penetration Capacity of Capped and Decapped Projectiles				
Target		Capped	Decapped	Decapped/Capped
Thickness	Obliquity			
lbs	deg.	fps	fps	%
200	60	980	1,090	111
200	65	1,030	1,210	117
240	60	1,225	1,320	108
Average (fps)		**1,078**	**1,207**	**112**

ADM 281/31 indicates the same tendency in cases of smaller shells attacking C armour plates at 30 deg.

    4) Angle of attack can be calculated as follows:

Let the tangent to the trajectory of the projectile at the point of impact and the plate normal line through the point of impact be represented by unit vectors in a three-dimensional Cartesian coordinate system. Now, calculate the angle between the two vectors.

    5) Finally, we apply Bu. Ord. Sk. 78841.

Mathematically:

1) $VL_\Sigma = \sqrt{\sum_{i=1}^{n} VL^2_i}$

2) $t_{eff} = t_{max} + \sum_{i=1}^{n} 0.7 t_i$

3) $VL \propto \dfrac{m}{m - m_c}$

4) $\vec{t} = \begin{bmatrix} 0 \\ \cos(\alpha) \\ \sin(\alpha) \end{bmatrix}$

$\vec{n} = \begin{bmatrix} \cos(\gamma)\sin(\beta) \\ \cos(\gamma)\cos(\beta) \\ \sin(\gamma) \end{bmatrix}$

$\Theta = \cos^{-1}\left(\dfrac{\vec{t} \times \vec{n}}{|\vec{t}| \times |\vec{n}|}\right)$

5) $VL = \dfrac{F\sqrt{td}}{41.57\sqrt{m}\cos(\Theta)}$

Whence
 VL=Ballistic Limit Velocity. {fps}
 VL_Σ=Total Ballistic Limit Velocity. {fps}
 m=Projectile Weight. {lbs}
 m_c=Armour Piercing Cap Weight. {lbs}
 t=Plate Thickness {in}, or Unit Vector (representing the tangent to the trajectory of the projectile at the point of impact).
 n=Unit Vector (representing the plate normal line through the point of impact).
 t_{max}=Thickness of Thickest Plate (laminated assemblies). {in}
 t_{eff}=Effective Thickness. {in}
 d=Projectile Diameter. {in}
 Θ=Impact Obliquity. {deg.}
 α=Angle of Descent. {deg.}
 β=Lateral Angle. {deg.}
 γ=Inclination of Armour Plate from the Vertical. {deg.}
 $F = 6*(t/d - 0.45)*(\Theta^2 + 2{,}000) + 40{,}000$.

Angle of descent and striking velocity values are extracted from range tables. If only abridged range tables are available, we interpolate the available angle of descent and striking velocity values then extract detailed values from the curves.

When noted, penetration values of Iowa are based on official penetration curves. These are illustrated below. When noted, penetration values of Yamato are based on the DeMarre formula adjusted to official data. For further details, see description below Figure 1.5.

Figure AB.1

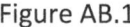

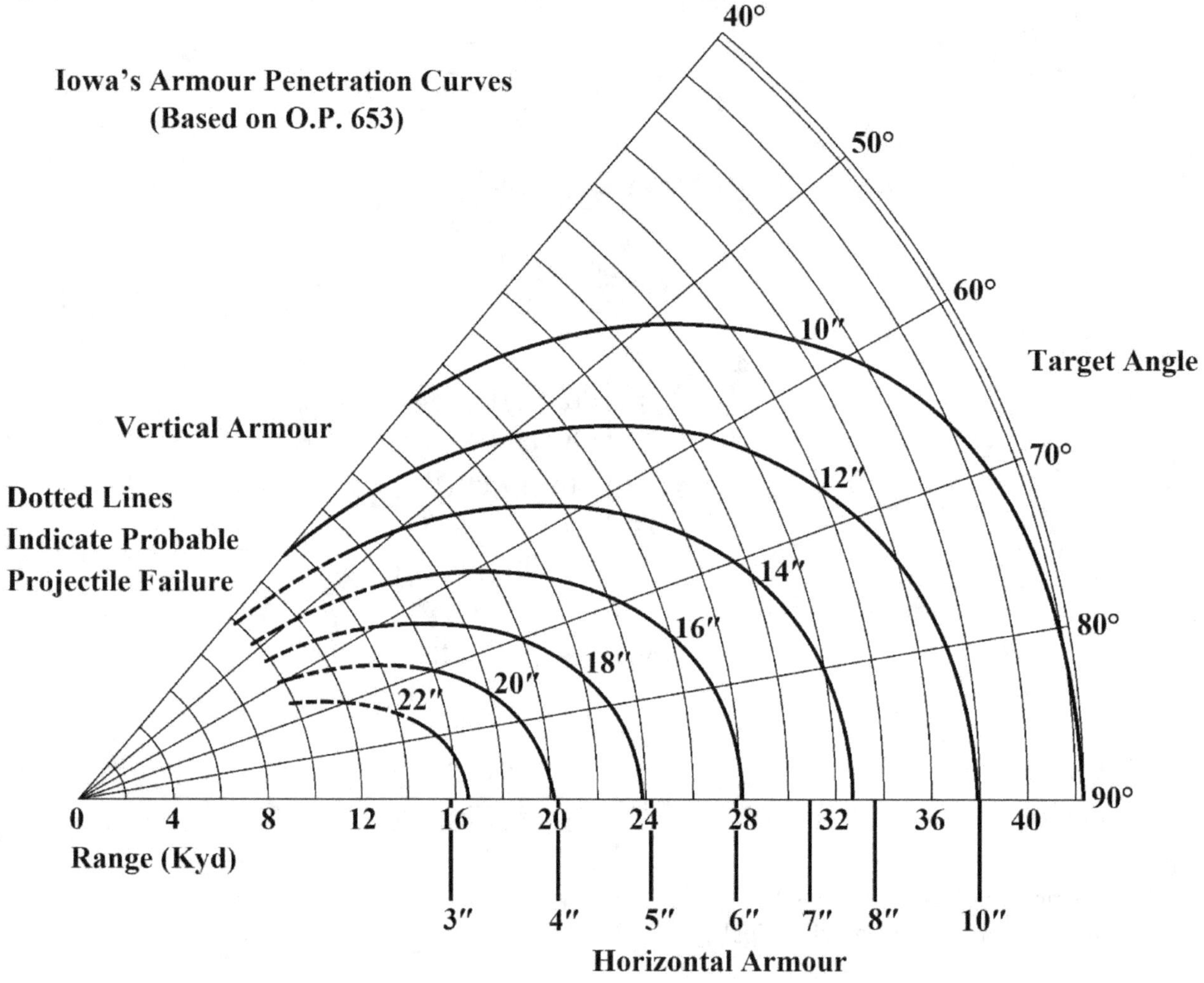
Iowa's Armour Penetration Curves
(Based on O.P. 653)
Vertical Armour
Dotted Lines
Indicate Probable
Projectile Failure
Target Angle
40°
50°
60°
70°
80°
90°
10″
12″
14″
16″
18″
20″
22″
0
4
8
12
16
20
24
28
32
36
40
Range (Kyd)
3″
4″
5″
6″
7″
8″
10″
Horizontal Armour

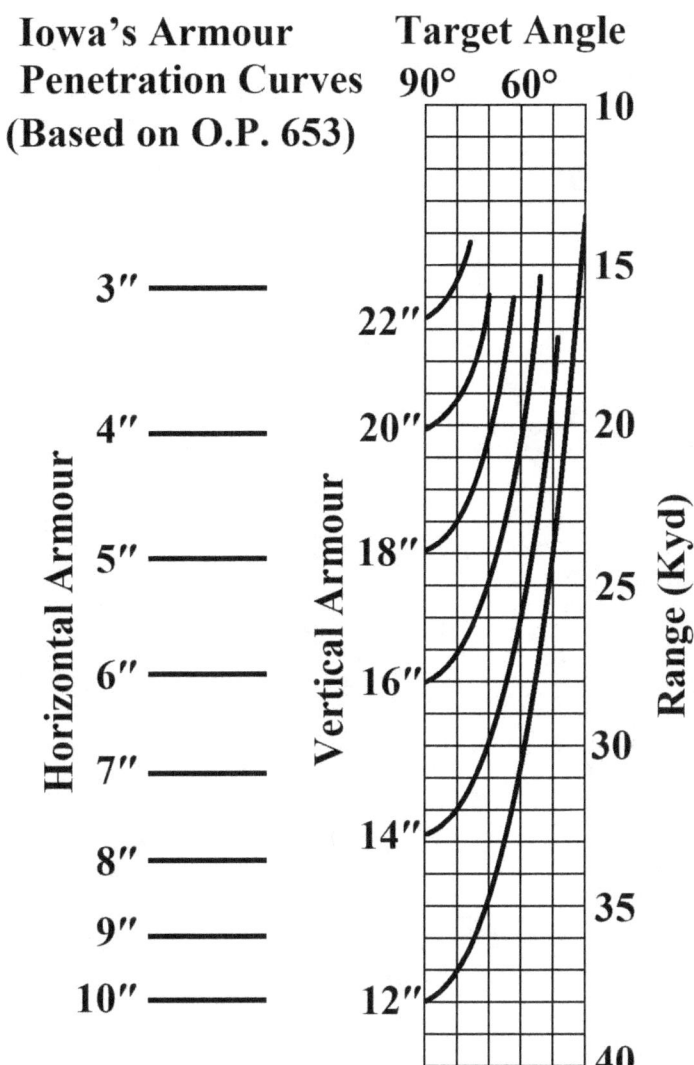

Appendix C – Japanese Armour Compared to Foreign Types

Table AC.1

JAPANESE ARMOUR COMPARISON WITH BRITISH ARMOUR Based on *Armour Technical Committee Meeting 20th November, 1947 Item 4. Japanese Armour.* (ADM 281/126) & *Armour Technical Committee Meeting 22nd July, 1948 Item 5. Japanese Armour.* (ADM 281/127)						
					Limit Velocity	
Thickness		Material	Test	Attack	Japanese /British Average	Japanese /British Best
HEAVY ARMOUR	600 lbs.	F.H./C.	Penetration	15in A.P.C. XVIIB/30°	1.09	1.06
	600 lbs.	F.H./C.	Perforation	15in A.P.C. XVIIB/30°	1.10	1.07
	480 lbs.	N.C./C.	Penetration	15in A.P.C. XVIIB/30°	0.97	0.92
	320 lbs.	N.C./C.	Penetration	12in A.P.C. VIIA/30°	0.97	---
	240 lbs.	N.C./N.C.	Penetration	15in A.P.C. XVIIB/60°	1.04	0.97
	240 lbs.	N.C./N.C.	Perforation	15in A.P.C. XVIIB/60°	1.12	1.05
AVERAGE	---	---	---	---	**1.05**	**1.01**
LIGHT ARMOUR	160 lbs.	N.C./N.C.	Penetration	12in A.P.C. VIIA/65°	0.93	0.92
	130 lbs.	N.C./N.C.	Penetration	8in S.A.P. IVB/60°	0.99	0.89
	120 lbs.	N.C./N.C.	Penetration	8in S.A.P. IVB/70°	0.88	0.86
AVERAGE	---	---	---	---	**0.93**	**0.89**
Note that the heaviest Japanese N.C. plates were attacked under conditions that are more favorable to C plates and their effective resistance was compared to those of C plates, diminishing their comparative performance. Also note that the first two tests were carried out against the same plate.						

Table AC.2

Performance of Japanese Armour Plates based on NPG 5-47				
Plate No.	Gauge	Class	Attack	B.L. given as % of Bu. Ord. Sk. 78841
JE-50-3133	7.25"	A	8" AP Mk. 21-3 at 30°	118 +/-1
JE-50-3133	7.25"	A	8" AP Mk. 21-5 at 30°	110-111
JE-50-3124	13"	A	14" AP Mk. 16-8 at 30°	87 +/-1
JE-50-3113	15"	A	14" AP Mk. 16-8 at 30°	82 +/-1
---	26"	A	16" AP Mk. 8-6 at 0°	90 +/-3
JE-50-3114	3.25"	B	6" AP Mk. 35-5 at 30°	104-105
JE-50-3116	3"	B	6" AP Mk. 35-5 at 30°	107 +/-1
JE-50-3120	3.25"	B	6" AP Mk. 35-5 at 30°	101-102
JE-50-3122	6"	B	8" AP Mk. 21-3 at 35°	98 +/-1
JE-50-3123	6"	B	8" AP Mk. 21-3 at 35°	98 +/-1
JE-50-3128	7"	B	8" AP Mk. 21-3 at 35°	94 +/-1
JE-50-3118	9"	B	12" AP Mk. 18-1 at 35°	94-95
JE-50-3108	12"	B	14" AP Mk. 16-8 at 30°	91 +/-1
Average Performance				**98.2**
Note that the first two tests were carried out against the same plate but the type of projectile was varied. Assuming a mean performance of the two impacts, the average performance of all plates drops to 96.8. However the heaviest Japanese B plates were attacked under conditions that are more favorable to A plates and their effective resistance was compared to those of A plates, diminishing their comparative performance.				

As can be seen, based on the U.S. test program, the average performance of the Japanese armour plates was 1.8 % below the predictions of Bu. Ord. Sk. 78841. If we assume that this is a good approximation of the mean performance of the armour plates carried by Yamato, then the ship's period of immunity against Iowa's guns would have been 17.7-34.0 kyd (16.2-31.1 km), not 17.0-34.5 kyd (15.5-31.5 km) – a miniscule difference.

If we base our calculation on the British test results and assume – as did the British, based on ADM documents – that British heavy armour was of comparable quality to U.S. armour, then Yamato's immunity would increase to 15.7-35.8 kyd (14.4-32.7 km) – a more substantial alteration.

Picture AC.1

26" (660 mm) armour plate, previously attacked by 2,700 lbs/16" (1,225 kg/406 mm) AP projectile at 90 deg. Limit velocity was established at 1,839 fps (561 mps) +/-3 %. Selfsame armour plates protected the frontal side of Yamato's main turrets, but they were sloped 45 deg. from the vertical. NPG 5-47 describes the actual turret face plates as impenetrable at any range due to the combination of extraordinary thickness and inclination. This conclusion is in agreement with our calculations.

Figure AC.1

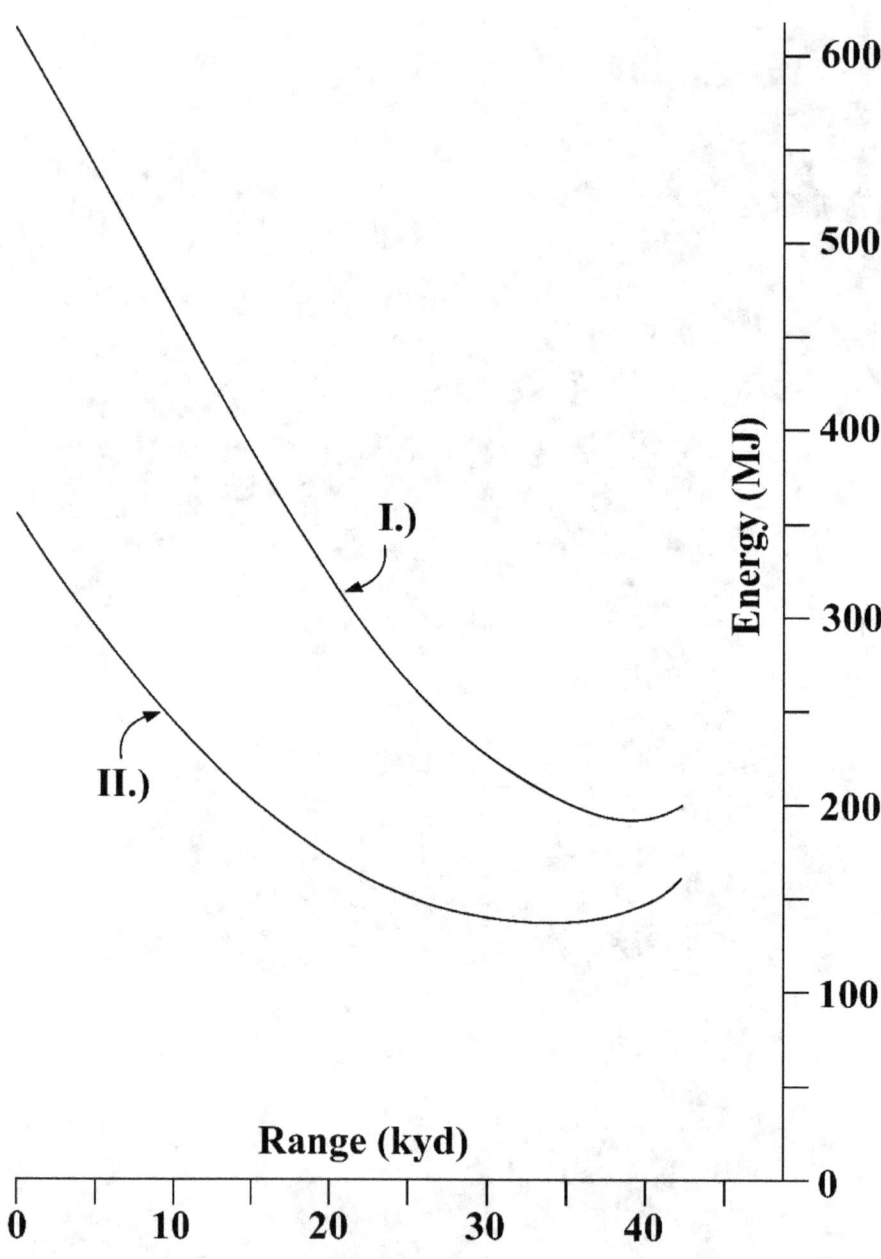

I.) Limit Energy of 26 in (660 mm)@45°
Turret Face Based on NPG 5-47
and Bu. Ord. Sk. 78841

II.) Terminal Energy of 16 in/2,700 lbs/2,500 fps
(406 mm/1,225 kg/762 mps) Projectile
Based on O.P. 770

Figure AC.2

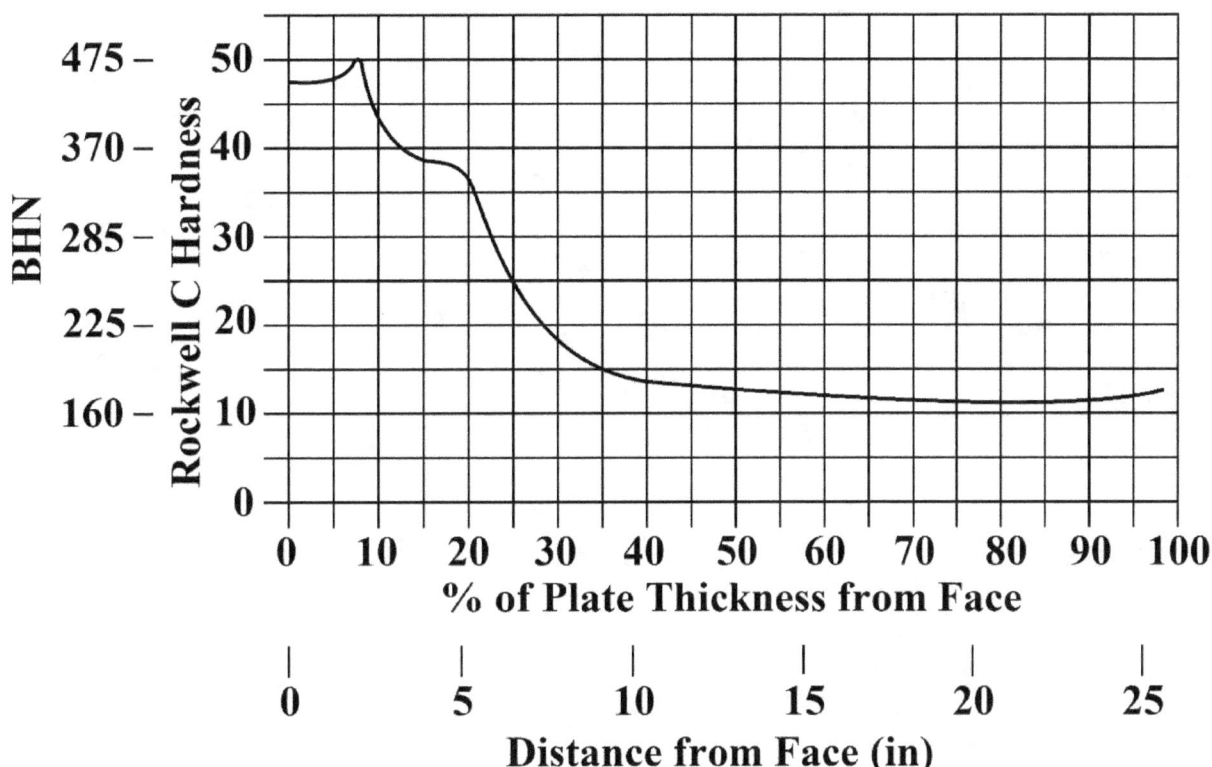

Appendix D – U.S. Armour

Table AD.1

Performance of U.S. Average Armour Based on NPG 5-47 and A.T.C. 22nd July 1948				
Source	Gauge	Class	Attack	B.L. given as % of Bu. Ord. Sk. 78841
NPG 5-47	7.25"	A	8" AP Mk. 21-3 at 30°	112.8
NPG 5-47	7.25"	A	8" AP Mk. 21-5 at 30°	109.7
NPG 5-47	13"	A	14" AP Mk. 16-8 at 30°	89.7
NPG 5-47	15"	A	14" AP Mk. 16-8 at 30°	89.6
NPG 5-47	3.25"	B	6" AP Mk. 35-5 at 30°	111.8
NPG 5-47	3"	B	6" AP Mk. 35-5 at 30°	110.6
NPG 5-47	6"	B	8" AP Mk. 21-3 at 35°	107.2
NPG 5-47	7"	B	8" AP Mk. 21-3 at 35°	105.4
NPG 5-47	9"	B	12" AP Mk. 18-1 at 35°	96.3
NPG 5-47	12"	A	14" AP Mk. 16-8 at 30°	95.1
A.T.C. 22nd July 1948	10"	A	14in A.P. (1500 lbs) 35°	87
A.T.C. 22nd July 1948	8.5"	A	12in A.P. (870 lbs) 35°	97
A.T.C. 22nd July 1948	9"	A	12in A.P. (870 lbs) 35°	97
A.T.C. 22nd July 1948	9.5"	A	12in A.P. (870 lbs) 35°	97
A.T.C. 22nd July 1948	10"	A	12in A.P. (870 lbs) 35°	97
A.T.C. 22nd July 1948	13.5"	A	14in A.P. (1500 lbs) 30°	90
A.T.C. 22nd July 1948	17"	A	14in A.P. (1500 lbs) 30°	90
A.T.C. 22nd July 1948	7"	B	8in A.P. (260 lbs) 35°	105.5
A.T.C. 22nd July 1948	10.7"	B	12in A.P. (870 lbs) 35°	97.2
A.T.C. 22nd July 1948	12"	B	14in A.P. (1500 lbs) 30°	95.1
A.T.C. 22nd July 1948	17.5"	B	14in A.P. (1500 lbs) 30°	91.6
Average Performance				**98.7**
Note that the first two tests were carried out against the same plate but the type of projectile was varied. Assuming a mean performance of the two impacts, the average performance of all plates drops to 98.1				

As can be seen, the mean performance of U.S. armour plates was also below the predictions of Bu. Ord. Sk. 78841 – by 1.3 % – and only 0.5 % above the mean performance of the Japanese plates.

Supposing that this is a good approximation of the mean ballistic efficiency of the armour plates carried by Iowa, the ship's period of immunity would have been slightly less than proposed above.

Appendix E – Flat Penetrators

<u>NPG report No. 4-44</u>

The shape of the projectile is an exceedingly important factor in determining its effectiveness at high obliquities, although this is a factor as yet only partially explored. To take the most extreme case, a flat-nosed projectile will penetrate a plate of quarter caliber thickness at an obliquity of 60° at velocity less than half that required for a projectile with a conventional 5/3 - caliber ogive…In this case the projectile neatly punches a clean hole in the plate, producing a negligible deformation in the material around the hole. As compared with more conventional projectiles, the reduction in the volume of metal plastically worked (because of the elimination of the dished area) accounts for the great reduction in the limit energy.

To take a less extreme case, a capped projectile with a broad cap having a sharp shoulder on it is more effective than a capped projectile with a small cap or with a rounded cap… Data with uncapped projectiles at high obliquities are unfortunately too meager to permit a full analysis of the effect of nose shape, but similar relations seem to hold – broad, angular noses "bite" better and result in lower limits at high obliquities than do the nose shapes usually used on uncapped projectiles.

Inspection of complete penetrations at high obliquity indicates that, as at low obliquity, the projectile experiences an initial yawing torque tending to increase the obliquity. The boundary between ricochet and penetration is probably determined by a delicate balance between the rate of yaw and the rate of deformation of the plate. A nose shape tending to decrease the rate of yaw in the initial stages of oblique impact would favor "biting" of the plate, and (other things being equal) should result in a lower limit velocity. To give the most extreme example again, the flat-nosed projectile very probably experience an initial righting torque – i.e., a torque tending to turn it towards the plate normal, rather than away from it.

Table AE.1

Summary of Test Results						
(NPG report No. 7-43)						
Projectile Weight (lbs)	Nose Shape	Obliquity (deg.)	T/D	Thickness (in)	Ballistic Limit Velocity (fps)	Penetration Limit Energy (%)
15	1.67 crh Ogive	0	0.24	0.73	589	100
11	Flat	0	0.24	0.73	663	91
15	1.67 crh Ogive	0	0.45	1.36	996	100
7.5	Flat	0	0.45	1.36	977	48
11	Flat	0	0.45	1.36	771	44
15	Flat	0	0.45	1.36	576	33
15	1.67 crh Ogive	30	0.45	1.36	1,023	100
11	Flat	30	0.45	1.36	905	58
15	1.67 crh Ogive	60	0.24	0.73	1,384	100
11	Flat	60	0.24	0.73	644	22

Table AE.2

Test Results					
(China Lake, California – December 1973)					
Half Nose Angle (deg.)	Weight of Projectile (grains)	Limit Velocity (fps)	Obliquity (deg.)	Plate Thickness (in)	T/D ratio
30 deg. (Sharp)	410	692	0	0.125	0.25
60 deg. (Blunt)	410	717	0	0.125	0.25
90 deg. (Flat)	410	651	0	0.125	0.25
30 deg. (Sharp)	410	782	30	0.125	0.25
60 deg. (Blunt)	410	787	30	0.125	0.25
90 deg. (Flat)	410	---	30	0.125	0.25
30 deg. (Sharp)	410	1410	60	0.125	0.25
60 deg. (Blunt)	410	---	60	0.125	0.25
90 deg. (Flat)	410	---	60	0.125	0.25
30 deg. (Sharp)	445	1290	0	0.25	0.5
60 deg. (Blunt)	450	901	0	0.25	0.5
90 deg. (Flat)	410	893	0	0.25	0.5
30 deg. (Sharp)	410	1422	30	0.25	0.5
60 deg. (Blunt)	450	1232	30	0.25	0.5
90 deg. (Flat)	410	---	30	0.25	0.5
30 deg. (Sharp)	410	---	60	0.25	0.5
60 deg. (Blunt)	445	2218	60	0.25	0.5
90 deg. (Flat)	410	---	60	0.25	0.5
30 deg. (Sharp)	410	1536	0	0.375	0.75
60 deg. (Blunt)	410	1193	0	0.375	0.75
90 deg. (Flat)	410	1297	0	0.375	0.75

Appendix F – Penetrator Hardness

<u>Effect of Projectile Nose Shape on Ballistic Limit Velocity, Residual Velocity, and Ricochet Obliquity.</u>

It became obvious from inspection of impacts and perforations made in the target plates that projectile rigidity had to be playing a major role in the differences being produced. The rigid (Rc 53), blunt (α = 90 deg) projectiles used in this research effort produced clean, square-bottomed impressions in the target plates at sublimit velocities. At velocities above the ballistic limit, plate perforations had the appearance of machined-like holes. Softer (Rc - 30) non-rigid fragments, for which the predictive equations were derived, deform upon impact and expend much more energy in projectile and plate deformation. To prove that the difference in experimental ballistic limit velocities as compared with predictions was attributable entirely to projectile rigidity, a test series using α = 90 deg, 410-grain projectiles having a hardness of Rc 30 was conducted against 0.250-inch steel plate at 0 deg obliquity. This test series produced a ballistic limit velocity of 1,698 ft/sec. The ballistic limit velocity determined for the Rc 53 fragment under exactly the same conditions was 893 ft/sec. Figure 5 illustrates the physical difference in the plate and projectile deformations associated with impacts and perforations by the two types of projectiles. The projectiles shown in the photograph are ones recovered after impacting at velocities just below the ballistic limit velocity. Note the amount of deformation experienced by the soft (Rc 30) projectile. The rigid (Rc 53) projectiles exhibited no deformation during any of the tests conducted during this research effort. Compare the plate deformations for both the incomplete and the complete perforations. The rigid projectile, by not deforming, maintains a concentrated shear stress gradient for promoting plugging. The non-rigid projectile, as it deforms, becomes a much less efficient penetrator and thereby must have a much higher impact velocity to achieve perforation… The results of this study illustrate the strong influence which penetrator deformation exerts upon the ballistic limit velocity of steel plates…It has been shown that a rigid projectile, to be considered as such, must truly remain rigid (exhibit almost no impact deformation) and that there can be a dramatic increase in ballistic limit velocity as projectile hardness decreases.

<u>NPG Report No. 4-44.</u>

A series of projectiles of hardnesses ranging from Rc 45 to Rc 65 were fired at fixed striking velocities and obliquities at a very thick plate of BHN 155. The harder projectiles passed through the plate; those of intermediate hardness were stopped inside the plate, but still headed through it, while the softer projectiles ricocheted after penetrating into the plate to various depths. The ricochet was evidently due to deformation of the projectile nose, which produced the same effect on the projectile as an increase in obliquity.

<u>NPG Report No. 3-47.</u>

…Work done at the Naval Proving Ground on AP projectiles equipped with caps of hardness greater than 600 Brinell (carbide ball) led to the development by the projectile manufacturers of 6" and 8" AP projectiles with considerably improved penetrative performance against Class A plate over standard projectiles. As a result of this work 6" and 8" AP projectiles are now equipped with caps of approximately 680 Brinell (carbide ball) as contrasted with the old production projectiles which had a cap hardness of 550 (carbide ball) Brinell.

Deeper hardness patterns and higher nose hardnesses for the bodies of AP and Common projectiles have been advocated by the Proving Ground for several years. It appears that the nose hardness should be as high as possible while the body hardness should be that which gives the maximum bend strength.

1) *Cap Hardness*

The present standard 12", 14" and 16" AP projectiles have caps of relatively low hardness 550 Brinell (carbide ball). As was pointed out in the introduction the penetrative ability of the 8" AP was increased by increasing the cap hardness so that increasing the cap hardness of the 12", 14" and 16" AP projectiles might increase their penetrative ability.

2) *Body Hardness*

Higher nose and body hardness for the subject projectiles might be desirable. With many steels the breaking strength on impact is greater when the steel is tempered at 350 °F than when the steel is tempered at 450 °F even though the steel is harder when tempered at 350 °F. Therefore, if the bodies of the subject projectiles were given a final tempering at 350 °F instead of 450-475 °F as at present, the ballistic performance might be improved. Tempering at 350 °F would give harder steel that is likely to have greater impact strength.

3) *Composition*

Most Navy projectiles regardless of size are made of steel of the same analysis (.60 C, 2.2 Cr and 3.2 Ni). Considering hardenability only, it may be that a more highly alloyed steel would produce better properties in the larger projectiles (12", 14" and 16") ...

Figure AF.1

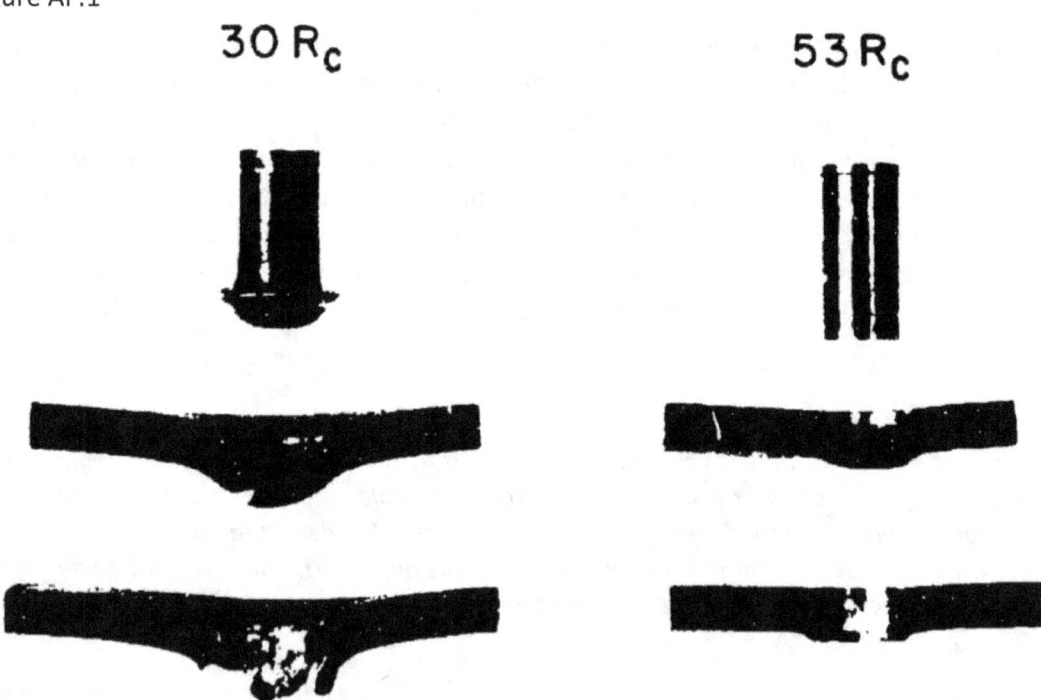

Effect of Penetrator Hardness on Projectile and Plate Deformation at Velocities Just Below and Above Ballistic Limit Velocity.

Appendix G – Radars

Table AG.1

Maximum range factors.
Wave length.
Size of target.
Height of target.
Target presentation (target angle).
Material of target.
Height of antenna.
Output power.
Sensitivity of receiver.
Atmospheric condition.
Type of indicator ("A" scope being the most sensitive).
Pulse repetition rate (determines maximum range scale that can be used).
Beam concentration.
Condition of radar equipment.
Operator's technique and skill.
Antenna size.
Pulse shape.
Antenna rotation rate.

Table AG.2

Height of Instrument + Height of Target		Visual Horizon		Radar Horizon	
ft	m	kyd	km	kyd	km
Surface Target					
100	30.5	21.6	19.7	24.9	22.8
150	45.7	26.4	24.1	30.5	27.9
200	61.0	30.5	27.9	35.2	32.2
250	76.2	34.1	31.2	39.4	36.0
Aerial Target					
1,000	305	68	62	79	72
2,000	610	96	88	111	102
3,000	914	118	108	136	125
4,000	1,219	136	125	157	144
5,000	1,524	152	139	176	161
6,000	1,829	167	153	193	176
7,000	2,134	180	165	208	190
8,000	2,438	193	176	223	204
9,000	2,743	204	187	236	216
10,000	3,048	216	197	249	228
20,000	6,096	305	279	352	322
30,000	9,144	373	341	431	394
40,000	12,192	431	394	498	455
50,000	15,240	482	441	557	509

Table AG.3

Characteristics of a Typical Fire-Control Radar	
Range)	
Maximum	Comparable to effective maximum range of guns
Minimum	Short
Accuracy)	
Range	Excellent
Bearing	Excellent
Resolution)	
Range	Excellent
Bearing	Excellent
Electromagnetic radiation frequency	High
Pulse repetition frequency	High
Wavelength	Short
Pulse width	Narrow
Scan	Predetermined Azimuth
Position	Atop/face of directors
Sensitivity to atmospheric conditions	High
Preferred bearing determination method	Lobe switching
Preferred electromagnetic radiation method	Continuous-wave or pulse-modulation

Table AG.4

Characteristics of a Typical Surface-Search Radar	
Range)	
Maximum	Slightly greater than LOS
Minimum	Short
Accuracy)	
Range	Good
Bearing	Good
Resolution)	
Range	Good
Bearing	Good
Electromagnetic radiation frequency	High
Pulse repetition frequency	High
Wavelength	Short
Pulse width	Short
Scan	360°
Position	Mast or Atop CT
Preferred bearing determination method	Maximum echo method
Preferred electromagnetic radiation method	Pulse-modulation
Antenna size	Relatively Small

Table AG.5

Characteristics of a Typical Air-Search Radar	
Range)	
Maximum	Greatest of any Radar
Minimum	Long
Accuracy)	
Range	Poor
Bearing	Poor
Resolution)	
Range	Poor
Bearing	Poor
Electromagnetic radiation frequency	Low
Pulse repetition frequency	Low
Wavelength	Long
Pulse width	Long
Scan	360°
Antenna	Bedspring
Antenna size	Large
Peak Power	High
Position	Atop Mast
Sensitivity to atmospheric conditions	Low
Preferred bearing determination method	Maximum echo method
Preferred electromagnetic radiation method	Pulse-modulation

Figure AG.1

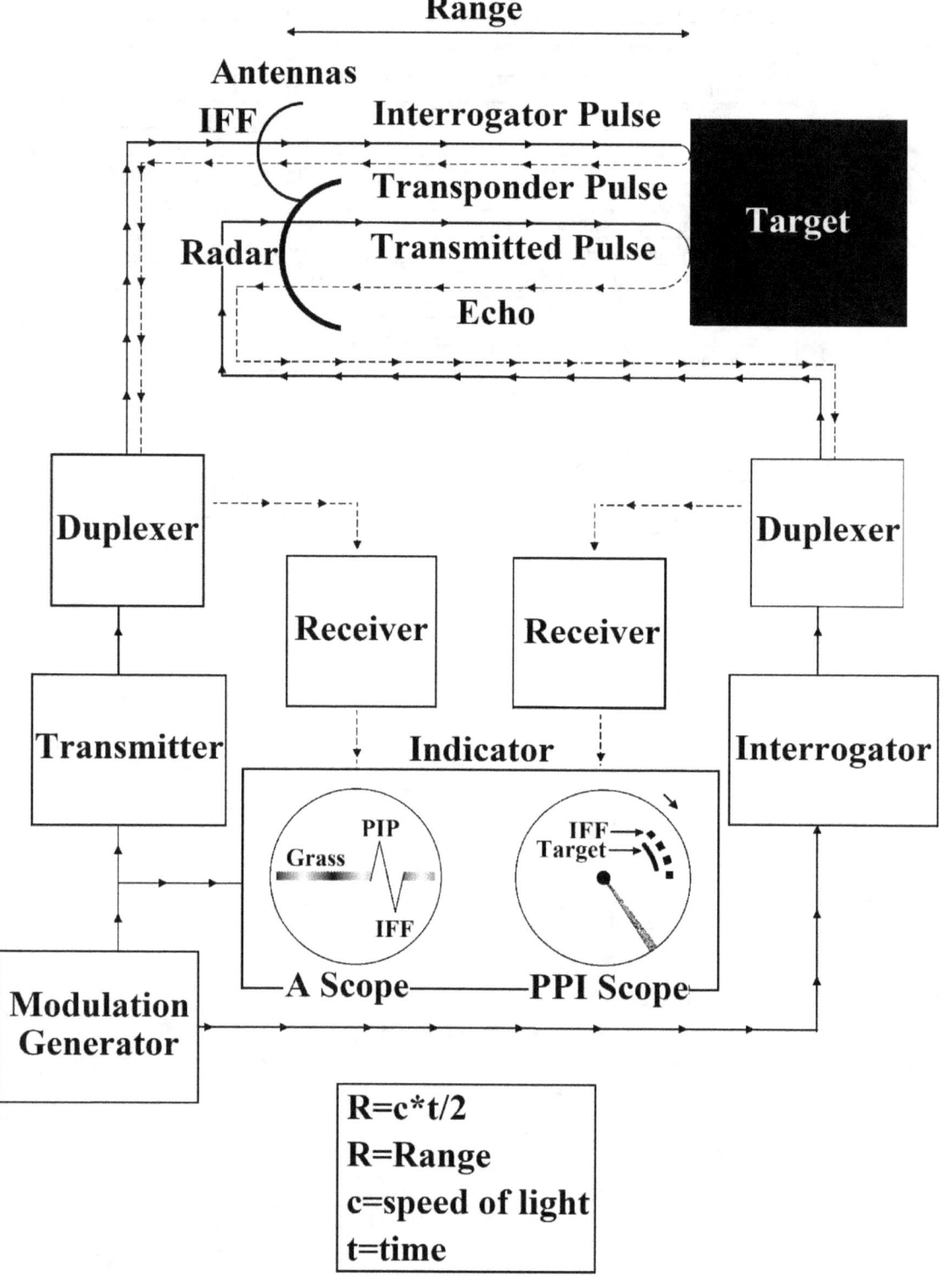

Figure AG.2

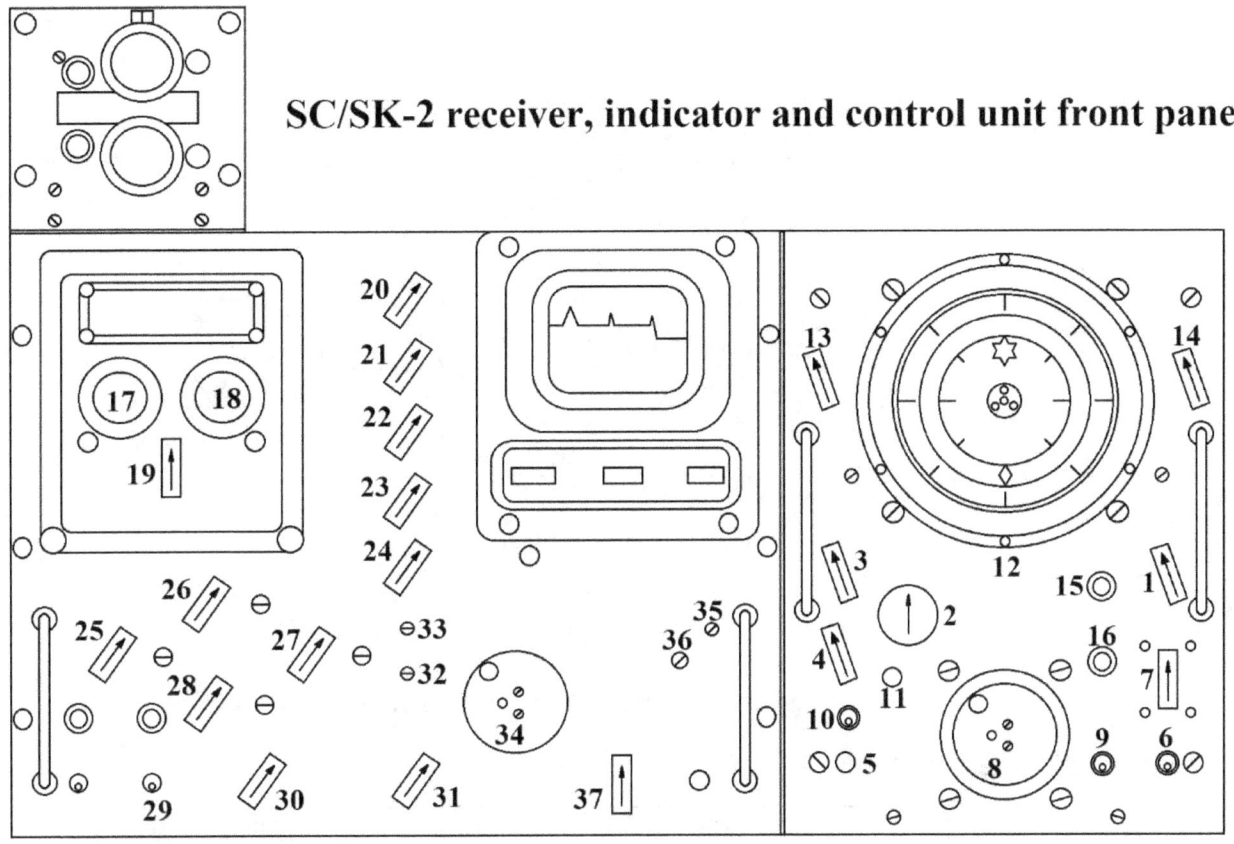

SC/SK-2 receiver, indicator and control unit front panel

SC/SK-2 Receiver, indicator and control		19	Receiver gain control
1	Main power switch	20	Receive-calibrate switch
2	Transmitter-plate voltage	21	Dial light brightness control
3	Relative-true bearing switch	22	Brilliance control
4	Remote bearing indicator switch	23	Focus control
5	Remote bearing mark	24	Astigmatism control
6	Automatic-manual toggle switch	25	IFF gain control
7	Antenna-control switch	26	Calibrate maximum
8	Hand crank for antenna control	27	Calibrate frequency
9	BL power switch	28	Calibrate minimum
10	Sweep	29	Challenge switch for IFF
11	Overload relay reset	30	Synchronizing switch
12	Bearing indicator	31	Crystal switch
13	Brightness control (bearing indicator light)	32	Range step height control
14	Brightness control (pilot lights)	33	Vertical trace centering control
15	Transmitter pilot light	34	Range crank
16	BL power pilot light	35	Horizontal sweep centering control
17	Radio frequency tuning control	36	Synchronizing pulse gain control
18	Local oscillator tuning control	37	Range selector switch

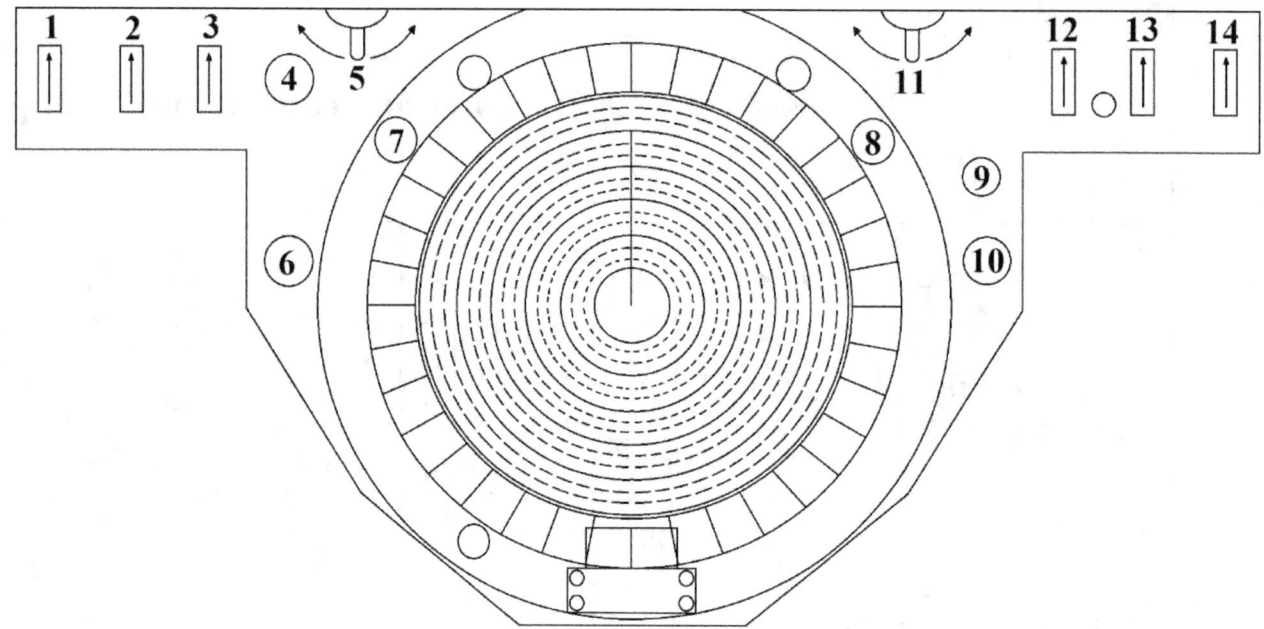

SC/SK-2. Plan position indicator (PPI)

SC/SK-2. Plan Position Indicator (PPI)	
1	Mark-IFF switch: normal operating position on IFF. When on MARK, range step is shown on PPI.
2	Dimmer control for PPI bearing dial light
3	PPI power switch.
4	Brilliance control.
5	Bearing indicator switch; RADAR-PPI: when on RADAR, bug follows the antenna; when on PPI, bug follows the yoke (cursor).
6	Focus control.
7	Bearing indicator adjustment control: for synchronizing bug reading and cursor reading, when operating bearing indicator switch is in the PPI position. Depress knob and set cursor (bearing blade) to read with the bug, then release knob to again engage the cursor.
8	Sector search control: in normal position, which is DOWN, clockwise rotation of the control increases the sector. Counterclockwise rotation narrows the sector. When pulled UP to engage the cursor, the sector may be rotated by rotating the cursor.
9	Remote alarm button.
10	Sector search off-on switch.
11	Relative-true switch for PPI.
12	Calibration control.
13	*Range selector switch*: Range 1-20 miles Range 2-75 miles Range 3-200 miles
14	Centering control: controls only centering of sweep along axis of sweep

SG Range and train indicator unit

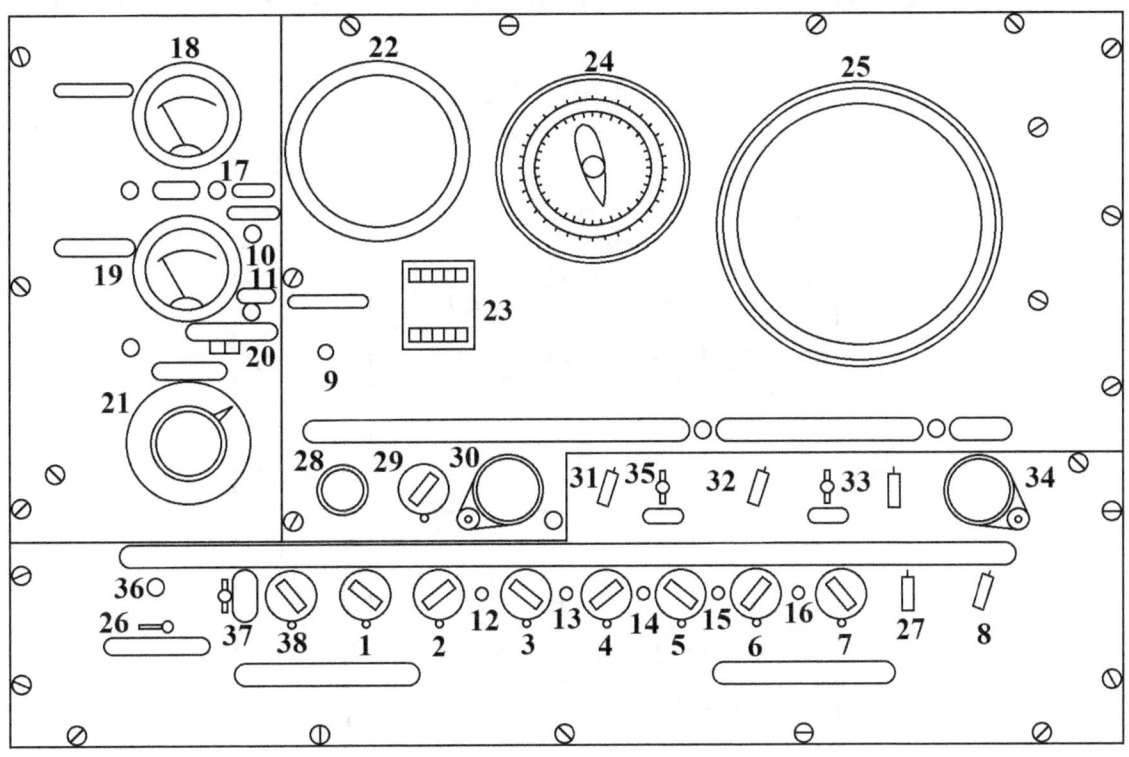

	SG – Range and train indicator unit.		
1	Range focus	20	Radiation switch
2	Zero set	21	Power supply control
3	Limit set	22	Range scope
4	Zero set	23	Range counters
5	Limit set	24	Bearing indicator
6	Pulse repetition frequency control	25	PPI
7	PPI focus	26	Synchro switch
8	Dial lights switch	27	When the radar is operating, switch K is in the NORMAL position
9	Pilot lights switch	28	Receiver gain control
10	Indicator F-902	29	Receiver tune control
11	Bearing Control F-901	30	Range crank
12	H center adjustment	31	Range scale switch
13	V center adjustment	32	Allows the operator to receive either signals or range markers on the range scope and PPI
14	Marker amplitude adjustment	33	Antenna's rotation control
15	PPI intensity adjustment	34	Remote range switch
16	PPI anode adjustment	35	Remote bearing switch
17	Remote control for the main-power switch at the transmitter-receiver unit	36	Reset button
18	Line voltage indicator	37	Determines the positions Off, Intermittent, and Continuous operation for IFF equipment
19	Transmitter current indicator	38	IFF gain adjustment

Mark 3 and Mark 4 Radar (FC, FD) Controls

Main Unit

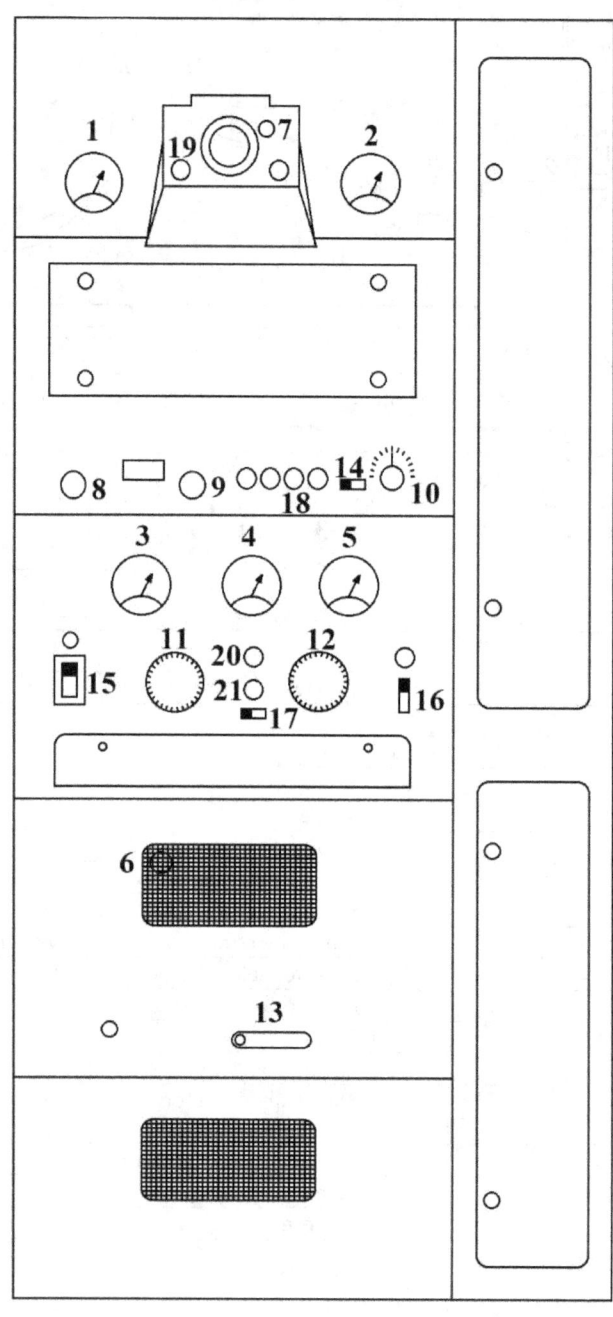

		MARK 3 AND MARK 4 RADAR (FC, FD) CONTROLS MAIN UNIT
	1	Plate current meter of modulation generator: should read about 200.
	2	Plate voltage meter of modulation generator: should read about 500.
	3	Load voltage: should be set to 120 at all times by means of control No. 11. (A recent directive says 115, but do not set it at 115 unless the set has been adjusted for this.)
	4	Magnetron plate current meter: should be set, to read about 30 by controls 13 and 12.
	5	Magnetron plate voltage meter: should be set to 12 (12,000 v.) by means of control No. 12.
	6	Magnetron filament voltage meter: should read 13.5. Can be seen by looking through the wire mesh on the front of the transmitter.
	7	Frequency control of modulation generator: adjusted by technician.
	8	Radio dial light dimmer: controls the brightness of the illuminated dial on the receiver.
	9	Receiver tuning control.
	10	Receiver sensitivity control:
	11	Load voltage control.
	12	Magnetron plate voltage control.
	13	Field control: adjusts plate current to the magnetron.
	14	Remote-local switch: determines whether the receiver sensitivity is controlled from the main unit by control No. 10, or whether the sensitivity is controlled by the receiver sensitivity knob on the range scope.
	15	Main off-on switch or line switch.
	16	Plate off-on switch.
	17	Dim-bright switch: controls brightness of the pilot lights on the face of the main unit.
	18	Mon jack: used in tuning up the receiver.
	19	Audio jack: used to obtain a synchronizing voltage when tuning up the receiver.
	20	Screw lock for 21.
	21	Magnetron filament voltage adjustment.

Range Unit

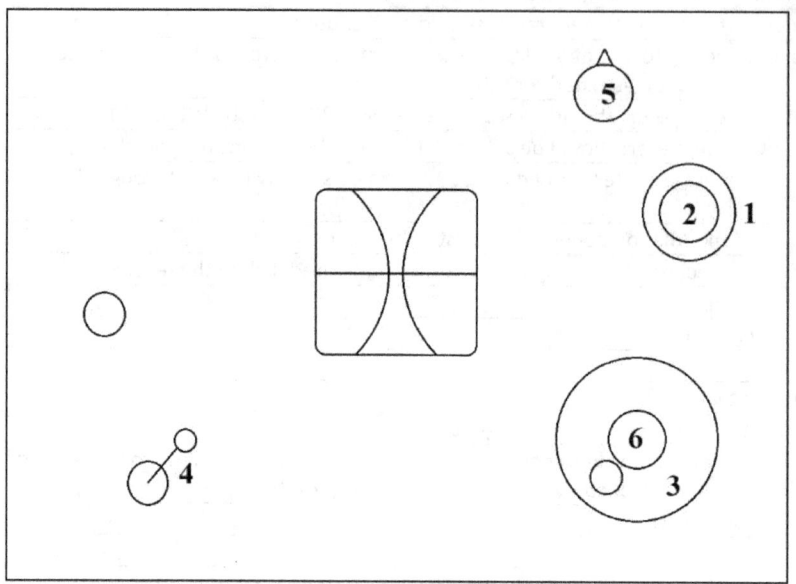

	MARK 3 AND MARK 4 RADAR (FC, FD) RANGE UNIT
1	Outer knurled nut: moves images across the scope.
2	Inner knurled nut: locks friction drive between the range knob, No. 3, and the electrical system controlling position of pips on the lace of the scope.
3	Range knob: moves images across the scope.
4	Pilot light bright-dim switch.
5	Dial light bright-dim control.
6	Signal button.

Control and Indicator Unit

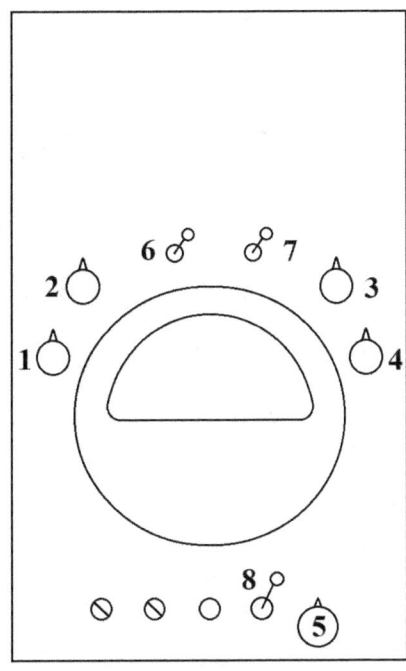

	MARK 3 AND MARK 4 RADAR (FC, FD) Control and indicator unit.
1	Intensity control: controls the brightness of the picture on the scope.
2	Image spread control: controls the size of the notch and expanded sweep.
3	Receiver sensitivity control: controls height of the grass and echoes.
4	Focus control: focuses the image on the face of the scope.
5	Sweep gain control: controls the length of the sweep. Should be completely clockwise.
6	Lobing on-off switch: turns lobing motor on or off.
7	Transmitter standby switch: turns the transmitter on or off. Used as a stand-by switch.
8	Pilot light dim-bright switch: (to be replaced by an A.G.C, switch.) Some sets have an anti-jamming switch above control

Train or Elevation Indicator

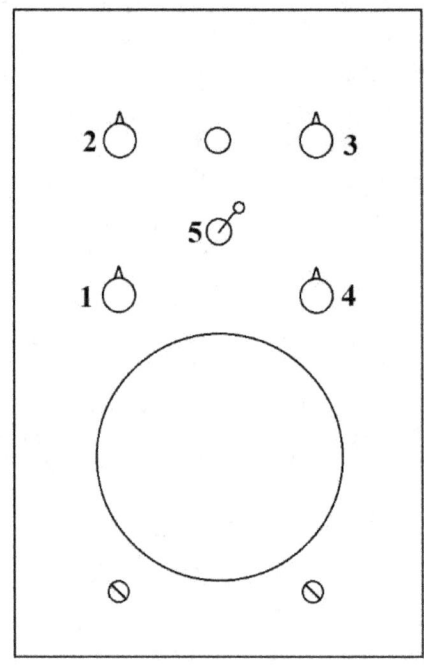

MARK 3 AND MARK 4 RADAR (FC, FD) Train or elevation indicator.	
1	Intensity control: controls the brightness of the image.
2	Image spacer control: move one sweep with relation to the other.
3	Sweep expansion control: opens or contracts the two steps.
4	Focus control: focuses the image.
5	Pilot bright-dim switch.

Table AG.6

Iowa's Air Search Radar (-1945)			
Set	SK (BB61,BB62,BB64) SK2 (BB63,BB62 from 1945)		
Maximum range (Antenna height=130 ft)			
BB, CV, CB, Large auxiliaries		51,500 yards	47,092 m
CA, CL, Medium auxiliaries		35,000 yards	32,004 m
Large planes		250,000 yards	228,600 m
Small planes		150,000 yards	137,160 m
Land		170 miles	274 km
Minimum range)			
"A" scope		1,500 yards	1,372 m
PPI 20-mile range		2.5 mi	4.2 km
PPI 75-mile range		6 mi	9.7 km
Resolution)			
Range		500 yards	472 m
Bearing		10 degrees	
Accuracy)			

Ranges	Sweeping		Stopped	
	Range	Bearing	Range	Bearing
30,000 yards	1,000 yds	3 degrees	200 yds	5 degrees
20 miles	1/2 mile	3 degrees	---	5 degrees
75 miles	1 mile	3 degrees	1-1/2 mile	5 degrees
200 miles	2 miles	3 degrees	---	5 degrees
375 miles	5 miles	3 degrees	1 mile	5 degrees

Wavelength	150 cm
Pulse Duration	5 microseconds
Power	200 kW
Antenna	17'x17' square 6'x6' dipole "mattress" array
Scan Rate	4.5 rotations per minute
Scope	A,PPI
Pulse Repetition Frequency	60 Hz
Frequency	200 MHz

Table AG.7

Iowa's Surface Search Radar (-1945)	
Set	SG
Maximum range)	
BB, CV, Large auxiliaries	35,000-45,000 yards (32,004-41,148 m)
CA, CL, Medium auxiliaries	28,000-35,000 yards (25,603-32,004 m)
DD, DM, AV, PC	18,000-30,000 yards (16,459-27,432 m)
Submarines	9,000-12,000 yards (8,230-10,973 m)
Submarine periscope	2,000-4,000 yards (1,829-3,658 m)
Large planes (altitude 1,000'-3,000')	20,000-35,000 yards (18,288-32,004 m)
Small planes (altitude 1,000'-3,000')	10,000-15,000 yards (9,144-13,716 m)
Minimum range)	
Ship	600 yards (549 m)
Plane	1,200 yards (1,097 m)
Accuracy)	
Range	±150 yds (137 m)
Bearing	±1 degree
Resolution)	
Range on the "A" scope	300 yards (274 m)
Range on the PPI	500 yards (457 m)
Bearing on the "A" scope	5 degrees
Bearing on the PPI	9 degrees
Wavelength	10 cm
Pulse Duration	2 microseconds
Power	50 kW
Antenna	48" by 15" (122 cm by 38 cm) cut parabola
Scan Rate	4, 8 or 12 rotations per minute
Scope	9" (23 cm) PPI / 5" (13 cm) A scope
Pulse Repetition Frequency	775, 800 or 825 Hz
Frequency	3000 MHz
Horizontal antenna beam width	5.6°
Vertical antenna beam width	15°

Table AG.8

Yamato's Air Search Radars		
Designation	Type 13	Type 21
Wavelength	2 m	1.5 m
Power	10 kW	25-30 kW
Range (Aircraft)	54.7 kyd (50 km)	76.6 kyd (70 km)
Range (Aircraft Group)	109.4 kyd (100 km)	131.2 kyd (120 km)
Range Accuracy	---	2.2-3.3 kyd (2-3 km)
Bearing Accuracy	---	10 deg.

Appendix H – Iowa's Mark 37 Fire-Control System

Figure AH.1

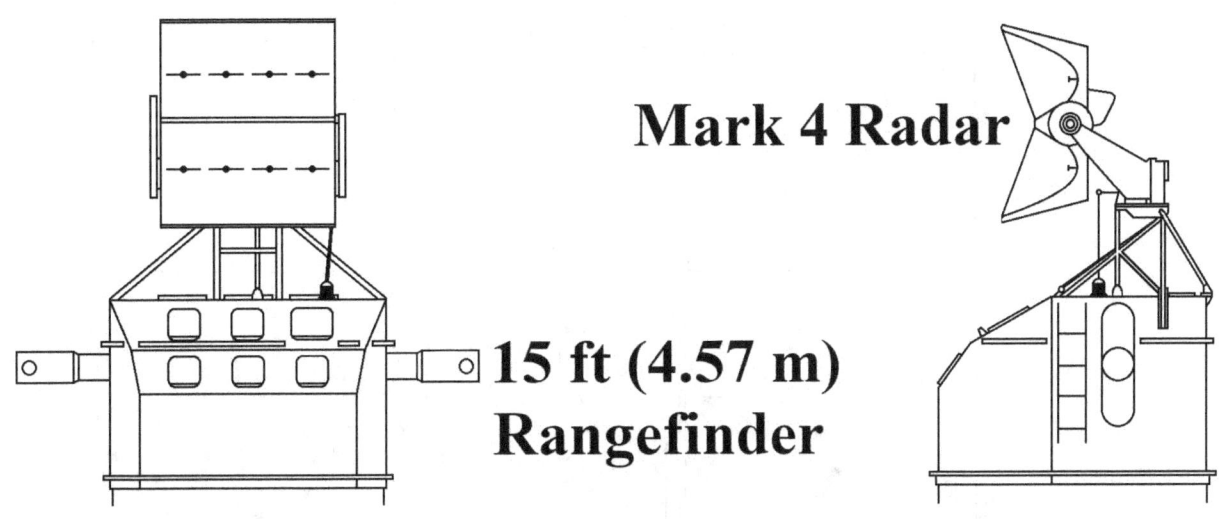

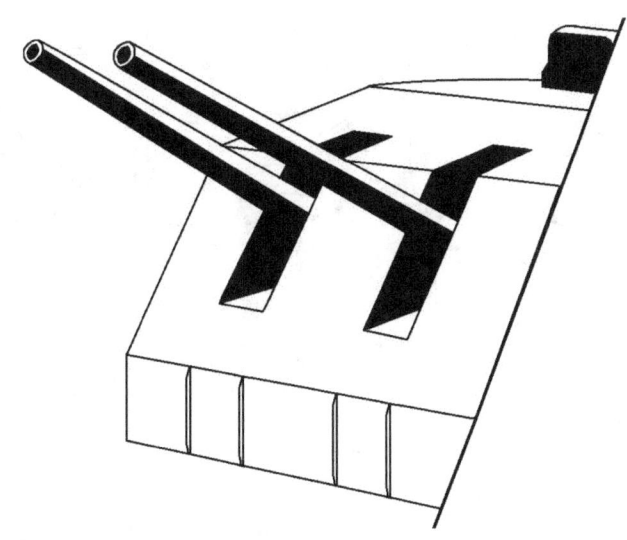

Mark 1 Computer

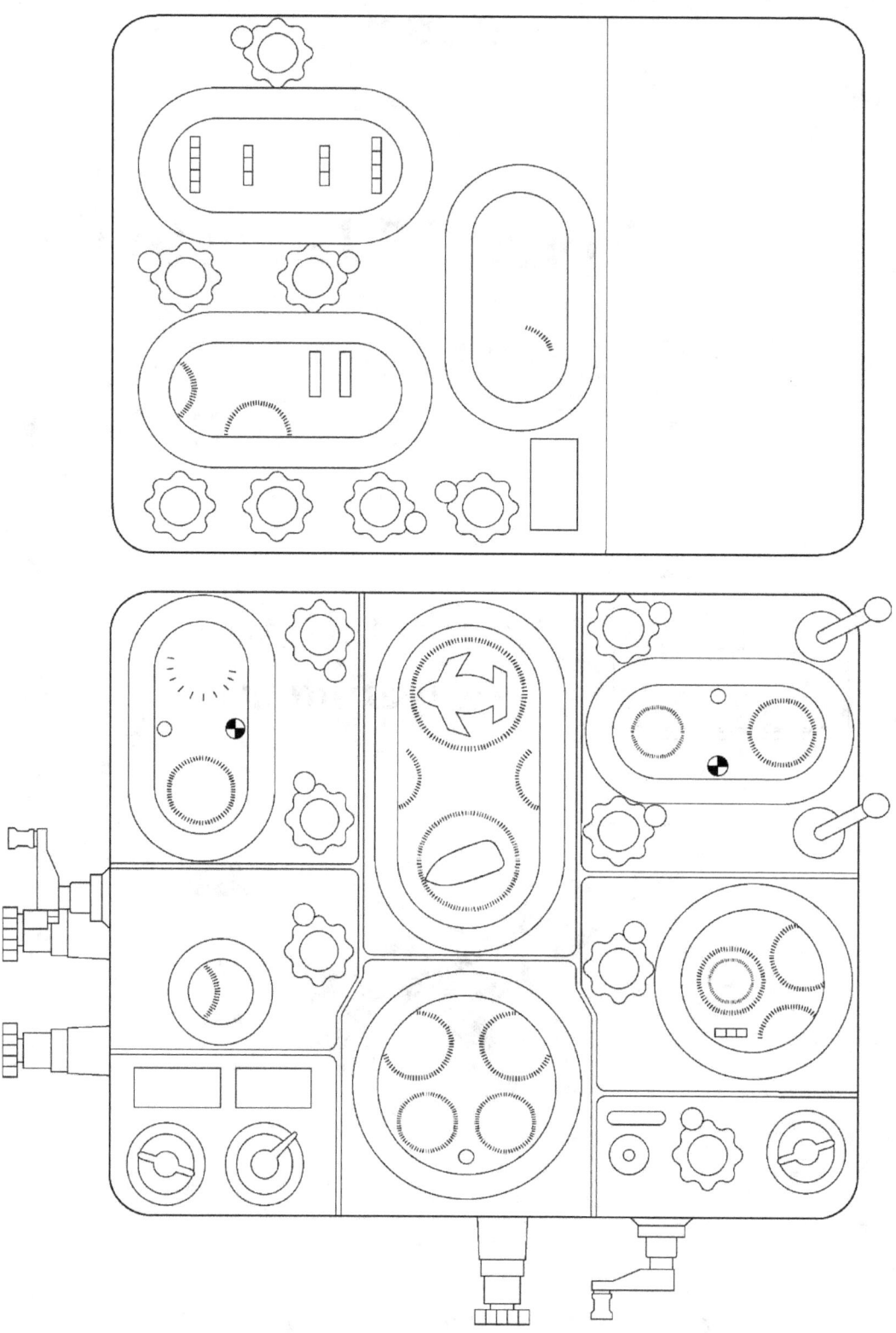

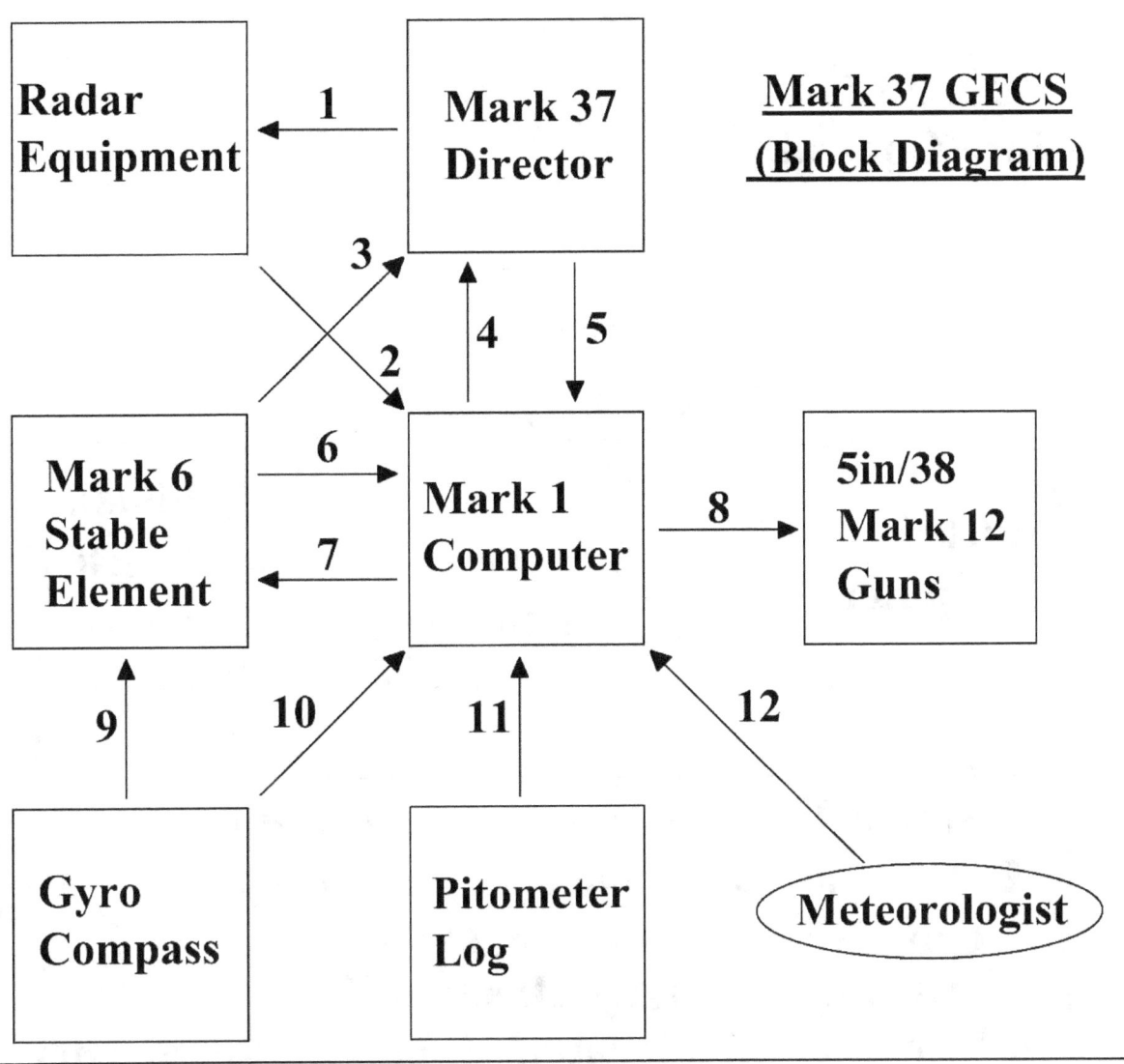

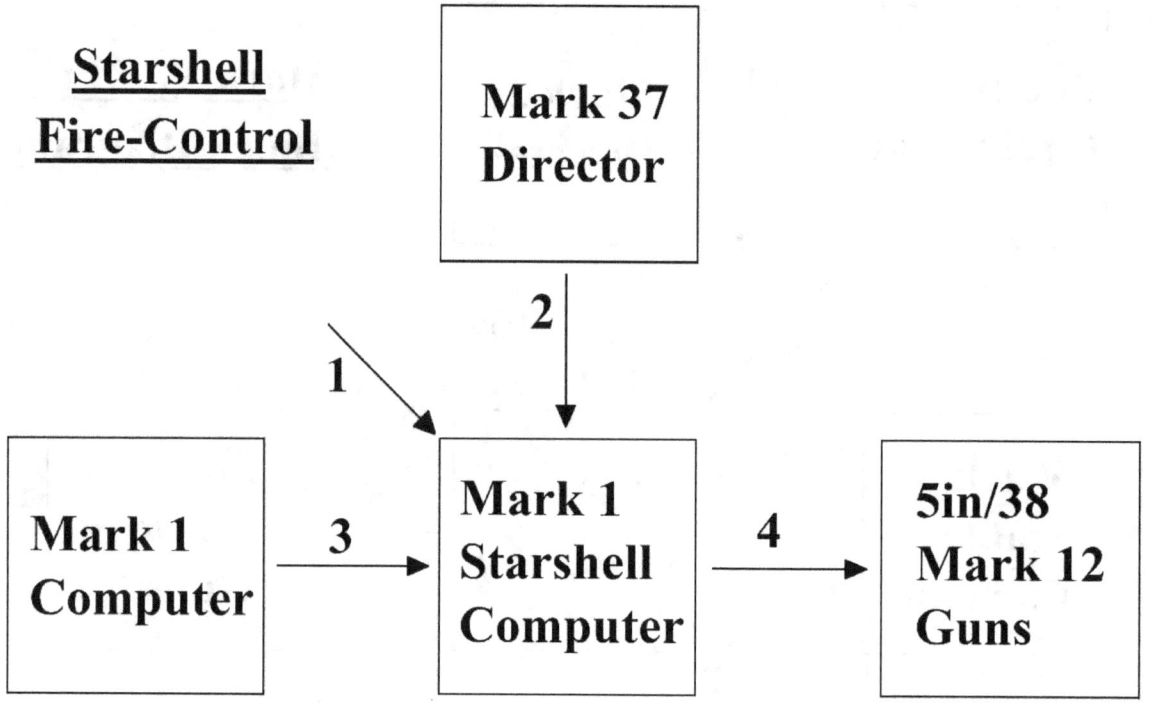

Appendix I – Yamato's Armour Decks

Figure AI.1

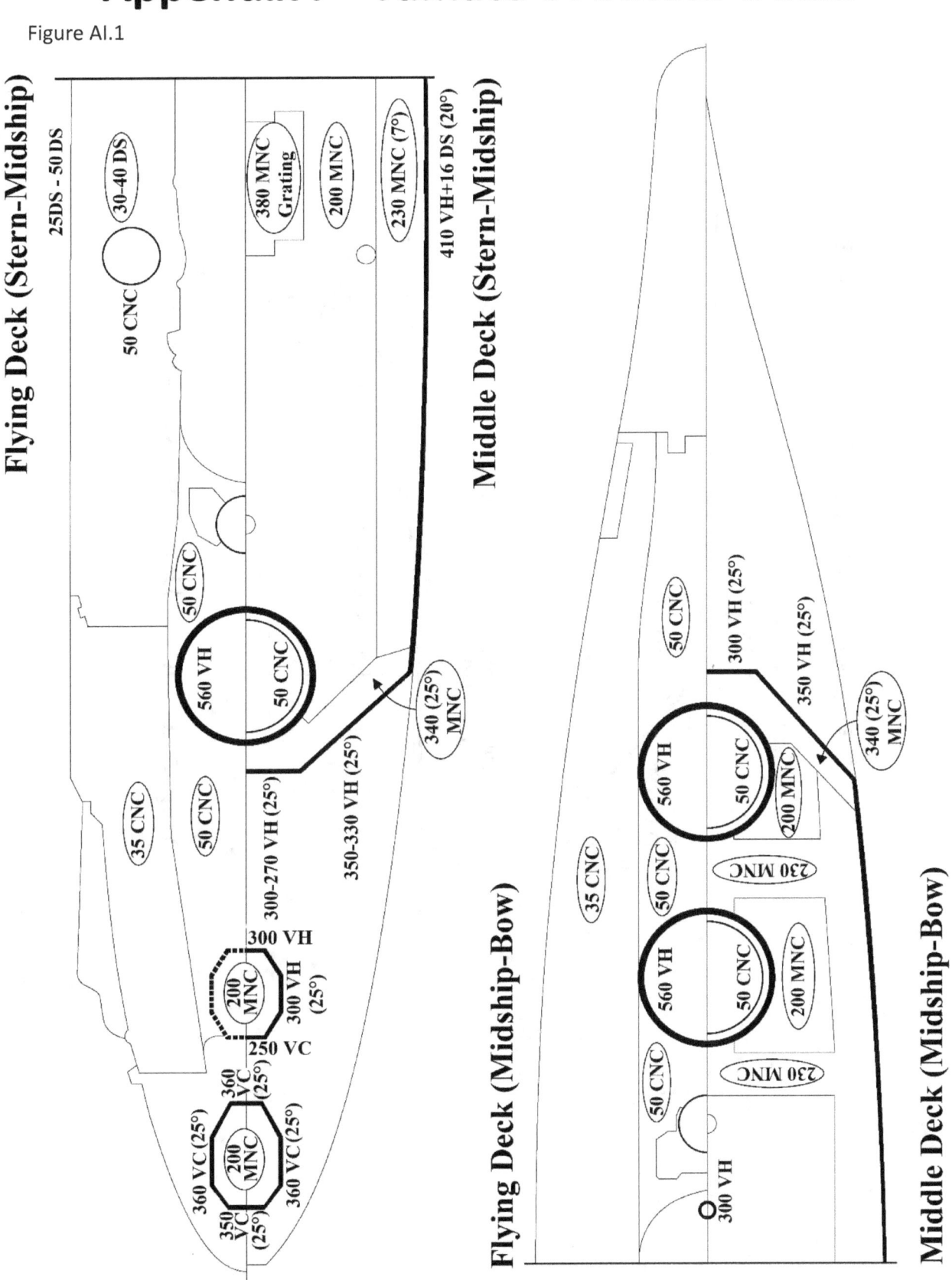

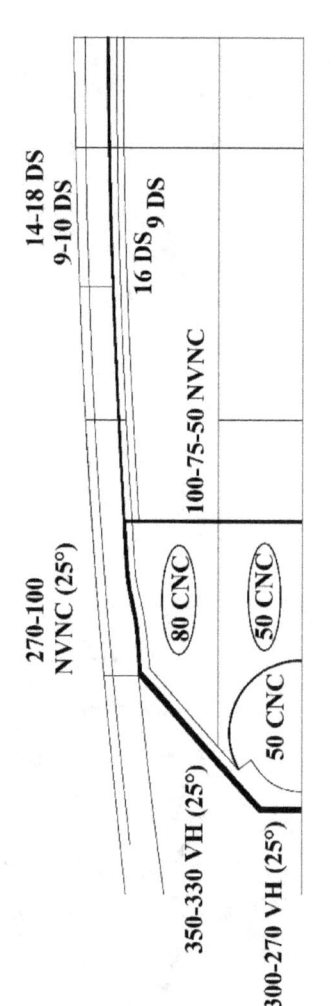

Magazine Bottom Deck (Stern–Midship)

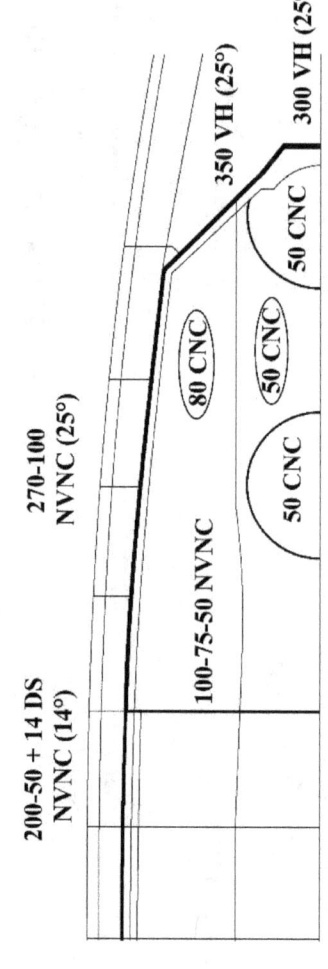

Magazine Bottom Deck (Midship–Bow)

Appendix J – U.S. Battleship's Combat Information Center (CIC)

Figure AJ.1
Battleship Combat Information Center (CIC)

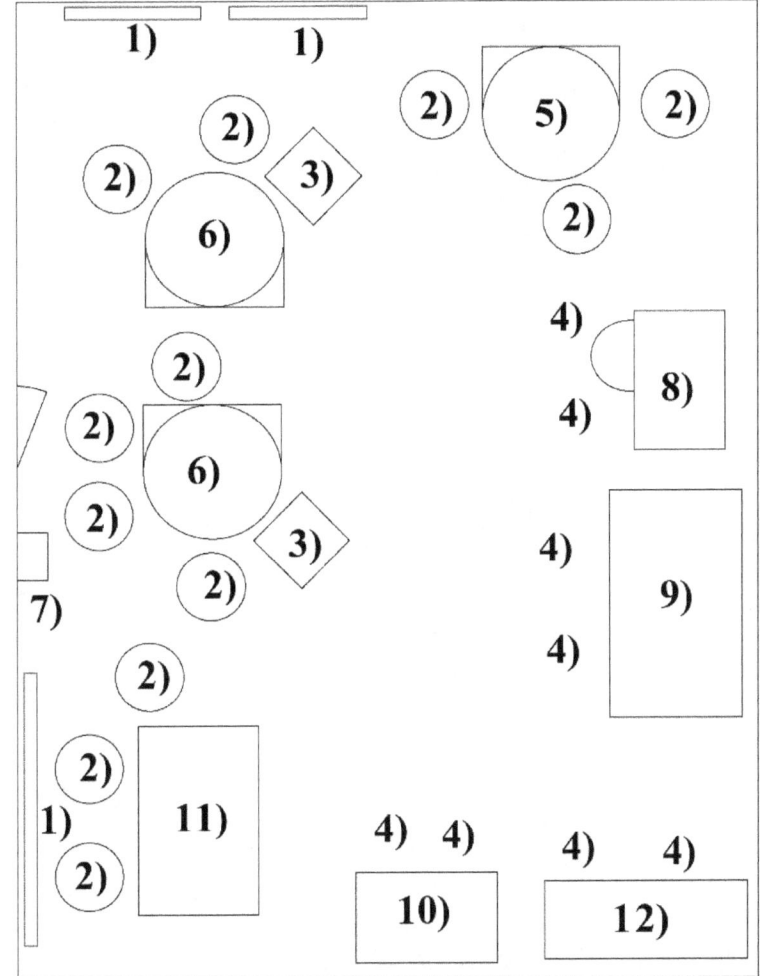

1) Status Board
2) Stool
3) Remote PPI
4) Chair
5) Summary Surface Plot
6) Intercept Plot
7) Speed Indicator
8) SK Receiver Indicator
9) SP Receiver Indicator
10) SG Receiver Train Indicator
11) Dead Reckoning Tracer
12) Radio Desk

Appendix K – Yamato's Main Turrets

Figure AK.1

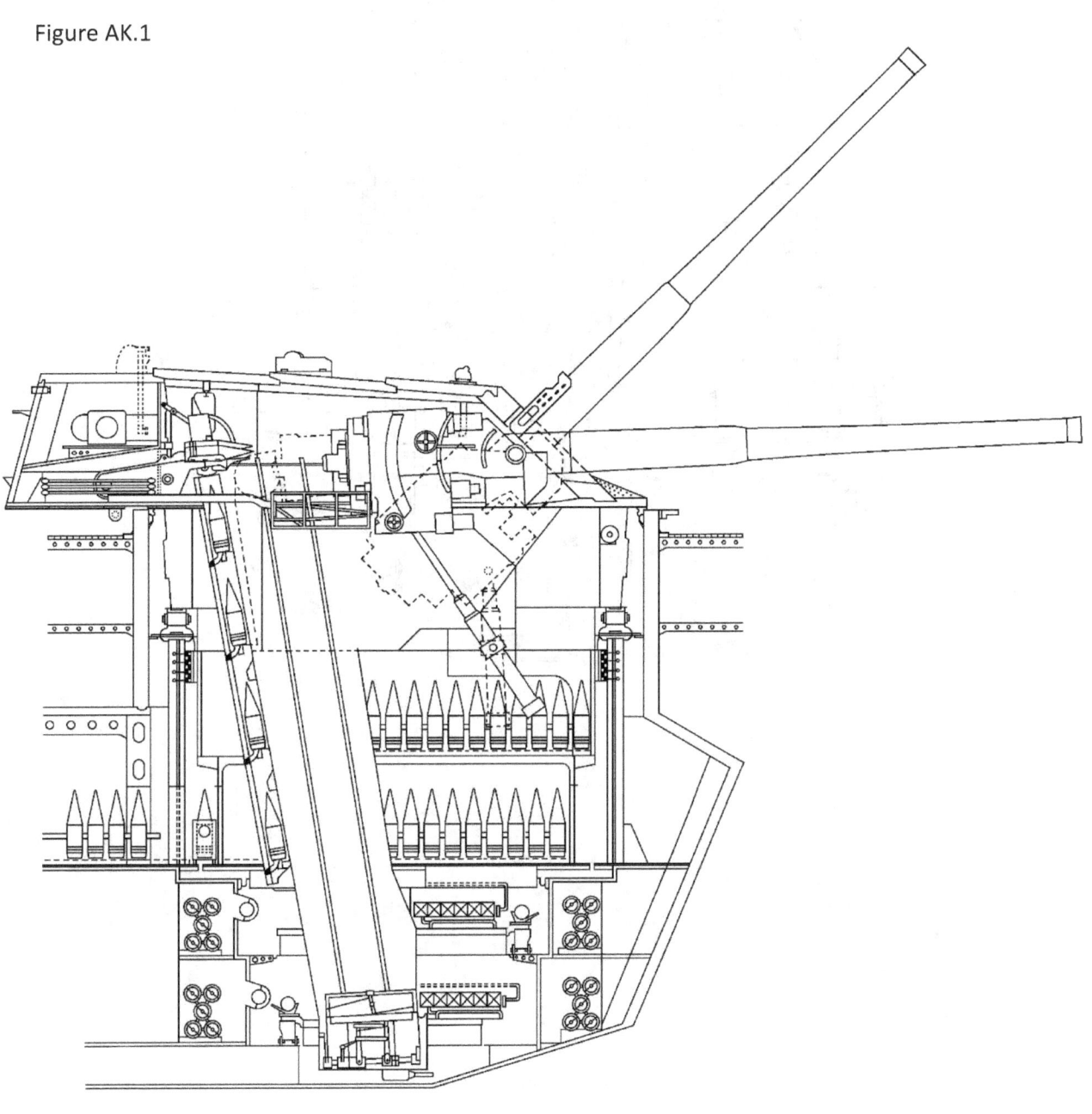

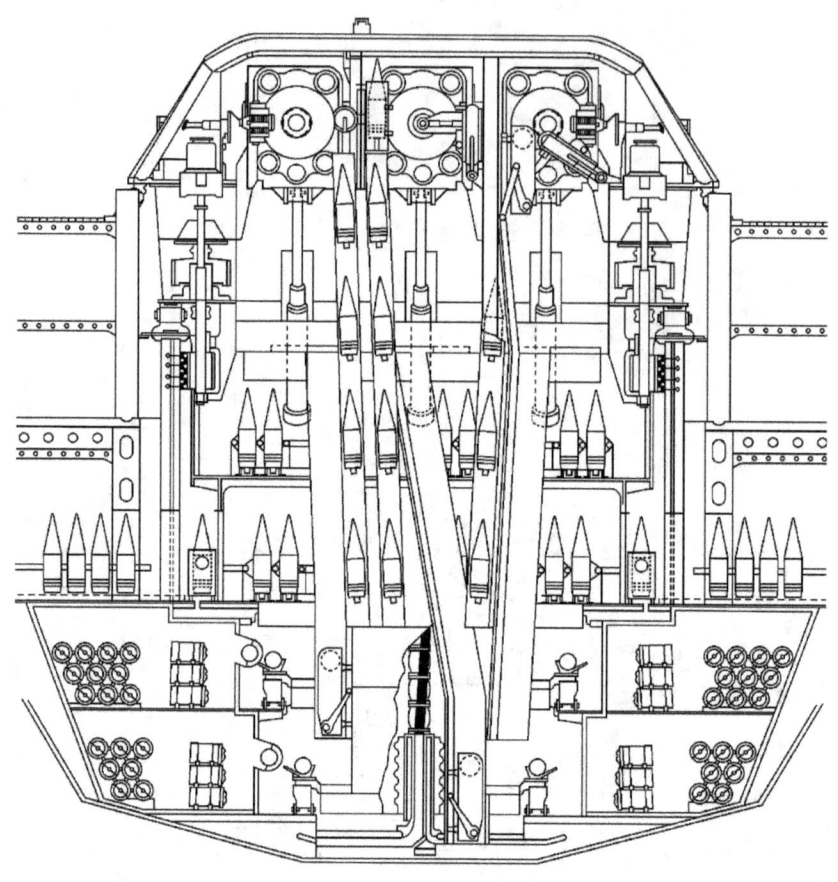

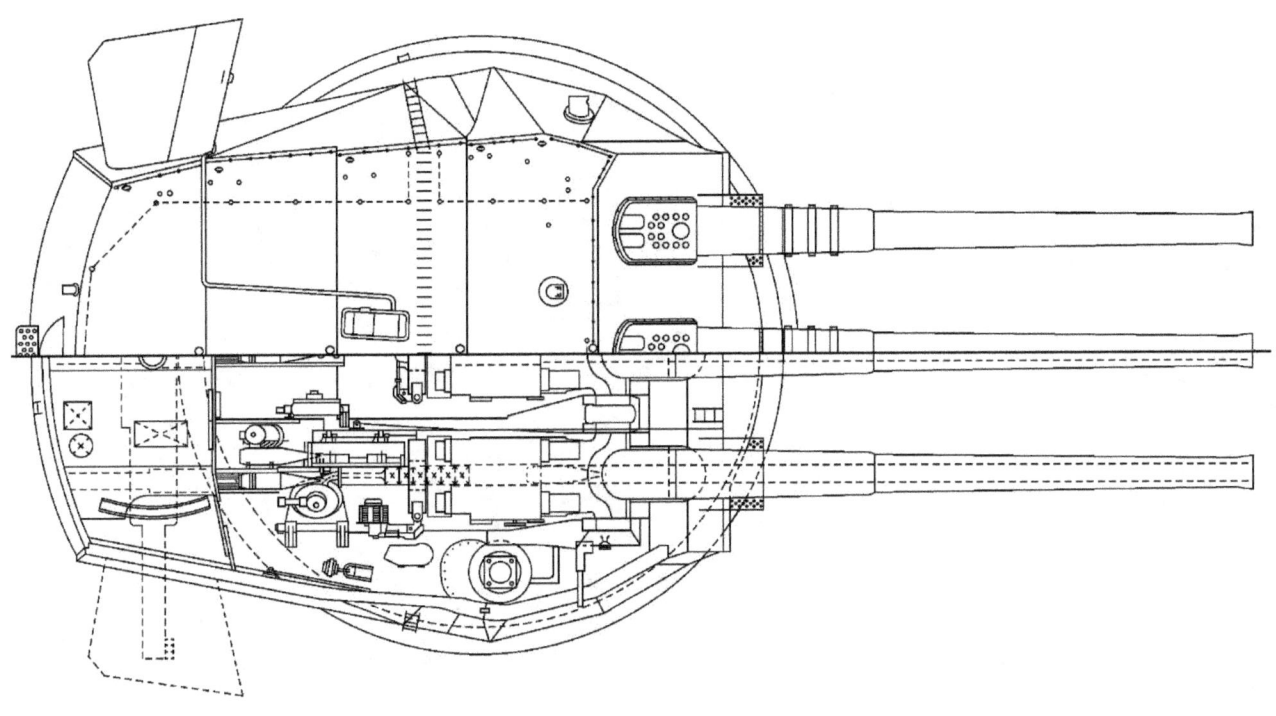

Appendix L – Projectiles

Figure AL.1

**HARDNESS DISTRIBUTION AND MACROSECTION
OF BETHLEHEM 16" AP PROJECTILE MK. 8-6 LOT 5 NO 6E9B
Brinell Hardness Values 10 mm Carbide Ball (3000 kg)-10 seconds
Etch. Ammonium Persulfate
Standard Projectile**

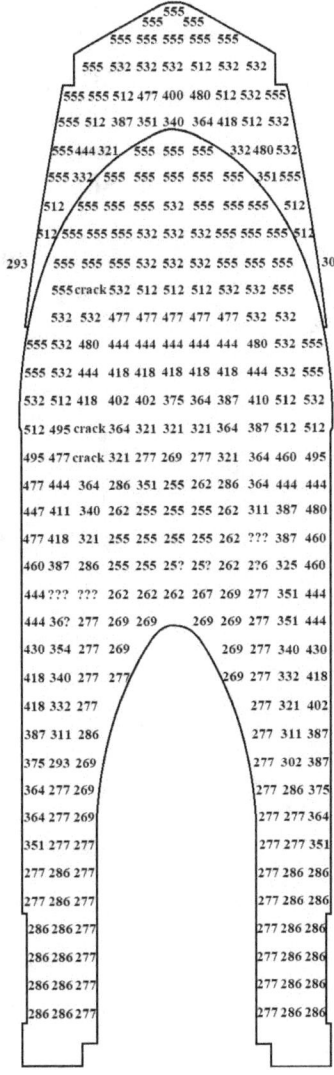

BETHLEHEM 16" AP PROJECTILE MK. 8-6 LOT 5 NO 6E9B

 **Cap**   **Body**

Cap	Body
Length: 6.8in (173 mm)	Length: 49.8in (1265 mm)
Weight: 312 lbs (141.52 kg) or 11.6%	Weight: 2191 lbs (993.82 kg) or 81.1%
Maximum Hardness: 555 BHN	Maximum Hardness: 555 BHN
Chemical Compositions)	Chemical Composition)
Carbon (C): .60%	Carbon (C): .60%
Manganese (Mn): .35%	Manganese (Mn): .34%
Phosphorus (P): .031%	Phosphorus (P): .025%
Sulfur (S): .018%	Sulfur (S): .017%
Silicon (Si): .39%	Silicon (Si): .26%
Nickel (Ni): 3.60%	Nickel (Ni): 3.33%
Chromium (Cr): 2.49%	Chromium (Cr): 2.30%
Molybdenum (Mo): .06%	Molybdenum (Mo): .09%
Copper (Cu): .10%	Copper (Cu): .10%

Windshield)

Length: 19.65in (499 mm)

Weight: 1.2%

U. S. NAVAL PROVING GROUND
DAHLGREN, VIRGINIA
17 Apr. 1947

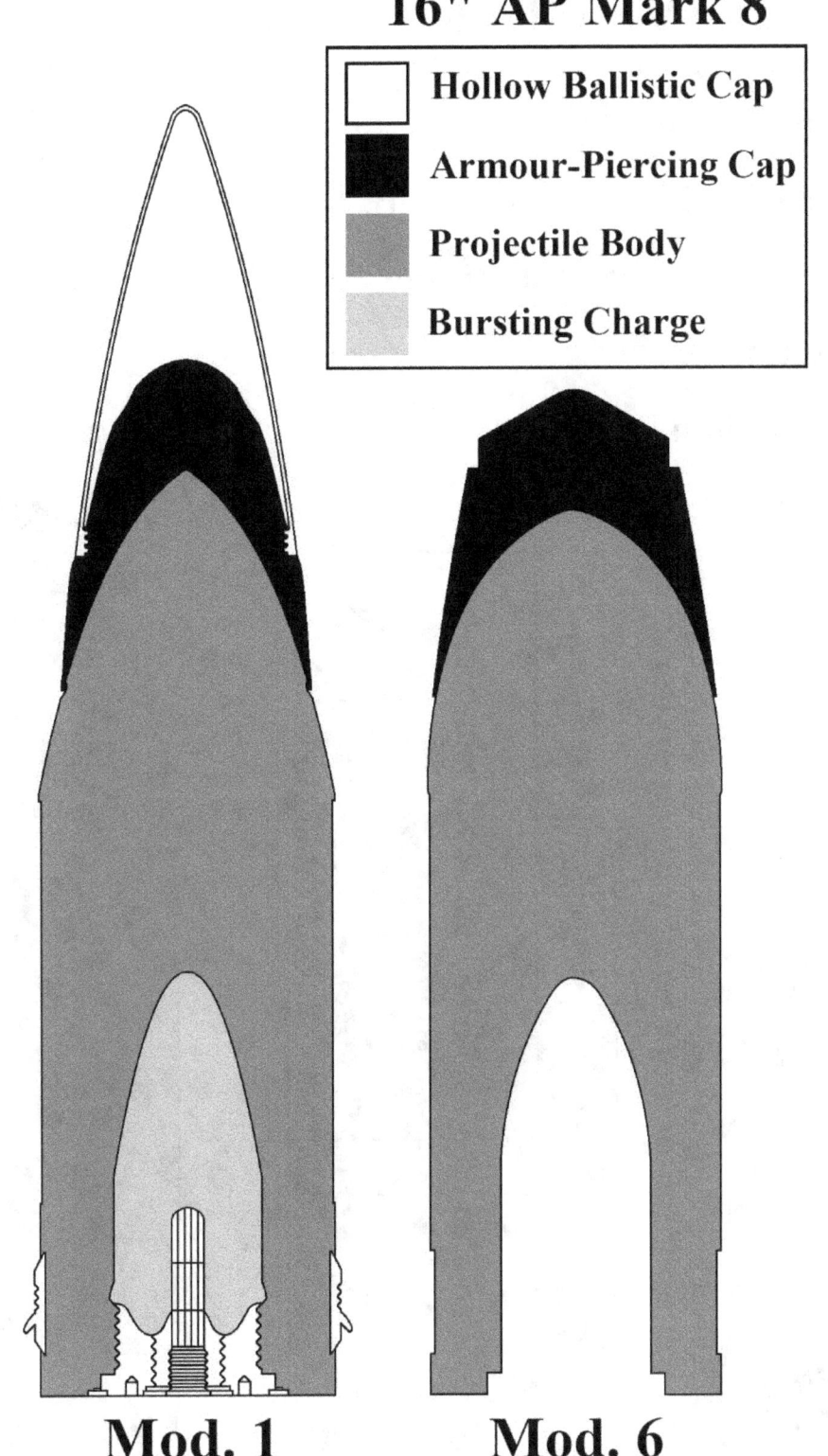

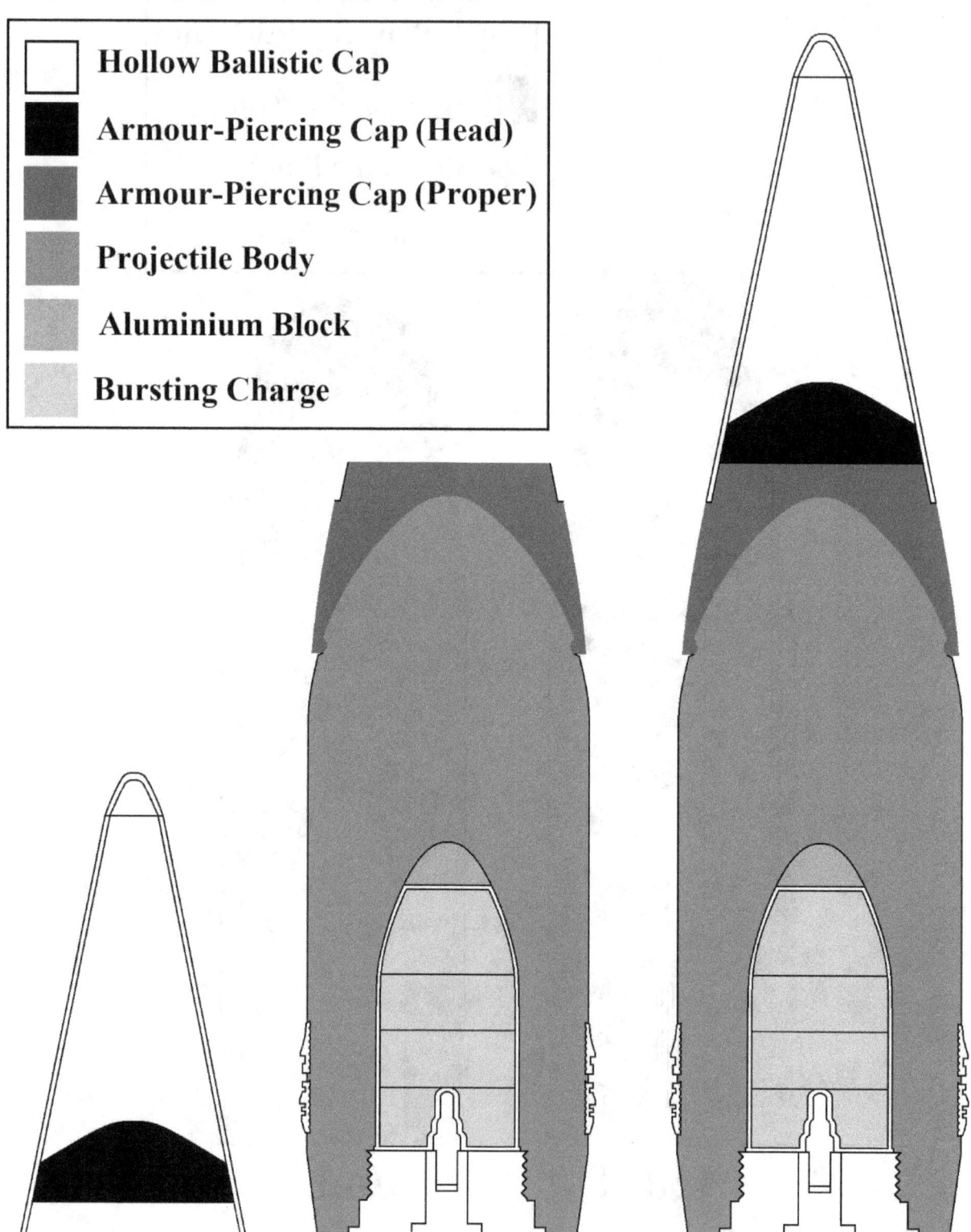

Figure AL.2
APC Type 91 (46 cm)

- Hollow Ballistic Cap
- Armour-Piercing Cap (Head)
- Armour-Piercing Cap (Proper)
- Projectile Body
- Aluminium Block
- Bursting Charge

Hardness Distribution (TMJ OT O-19)

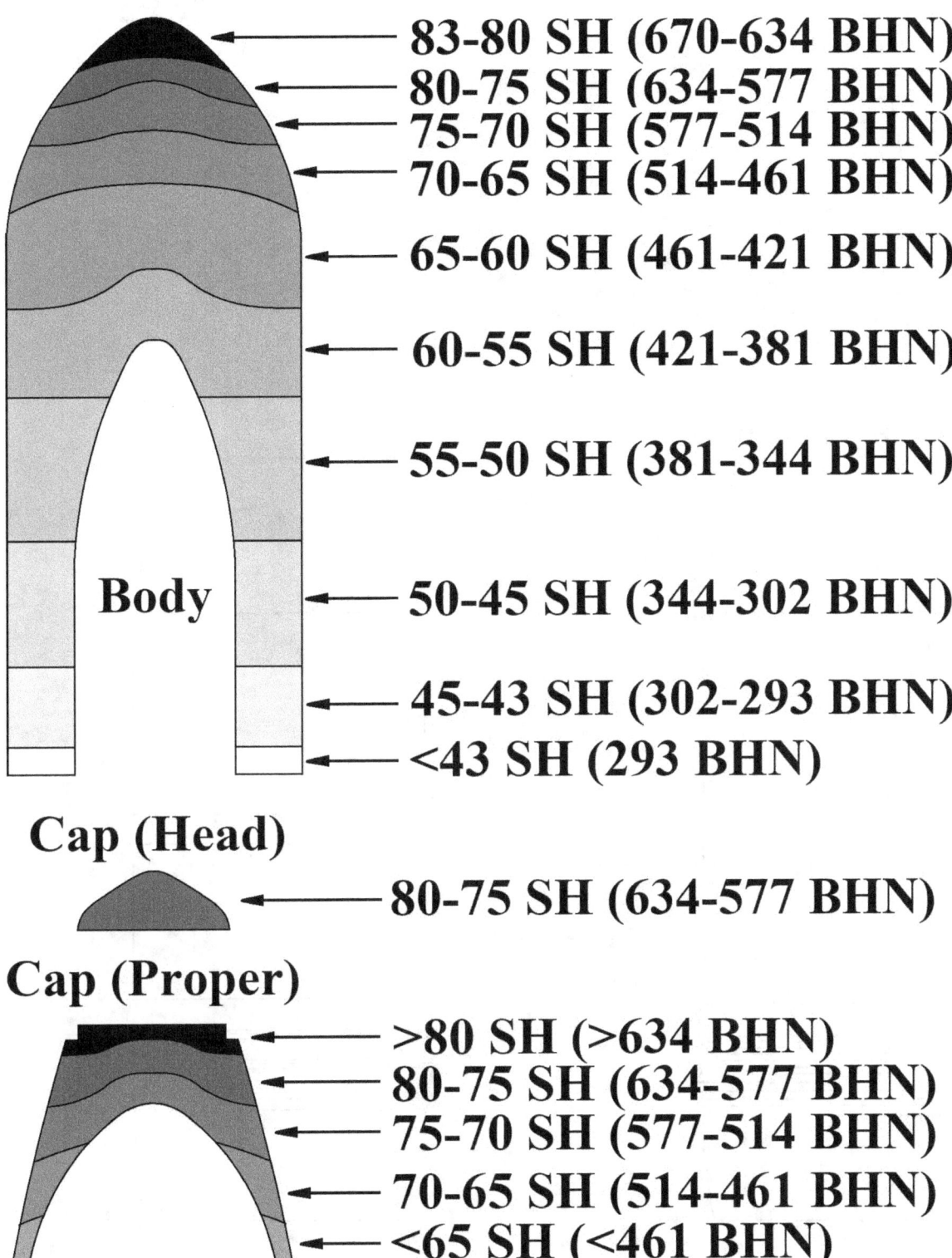

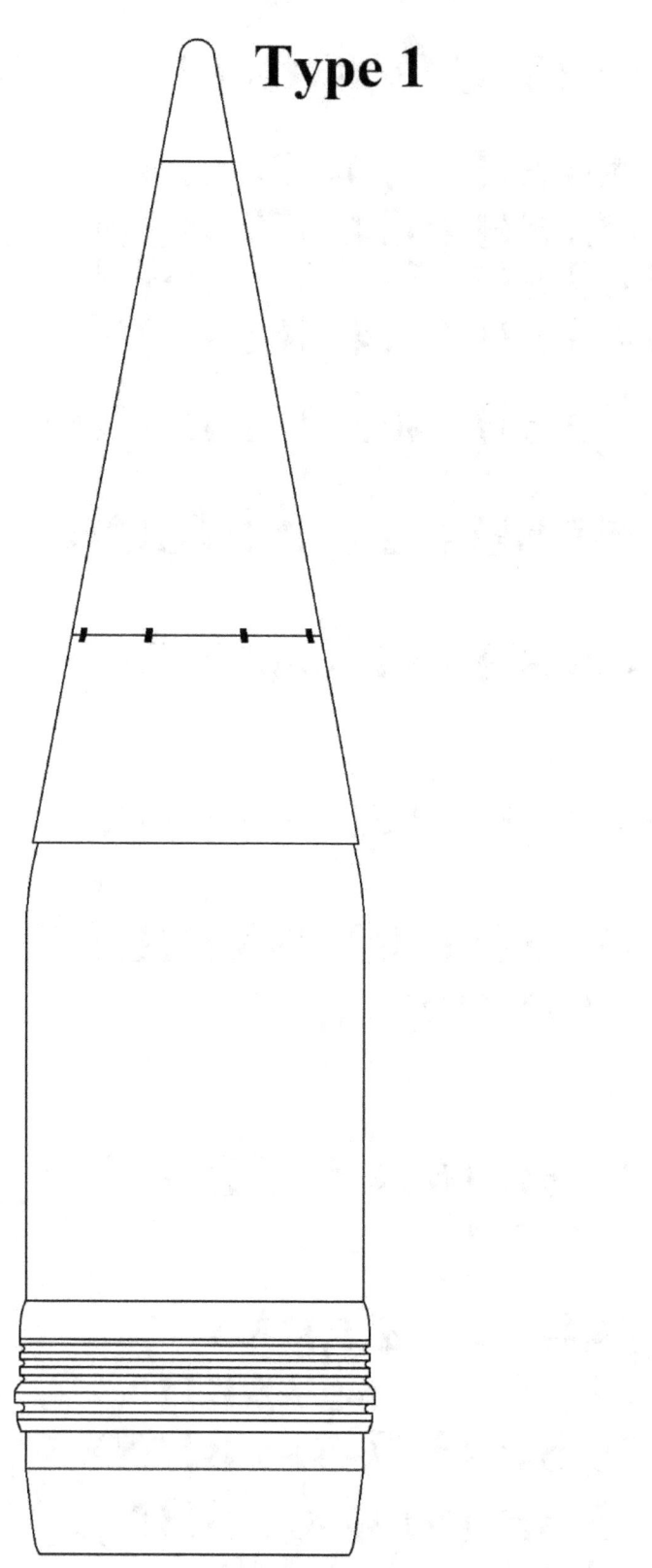 Type 1

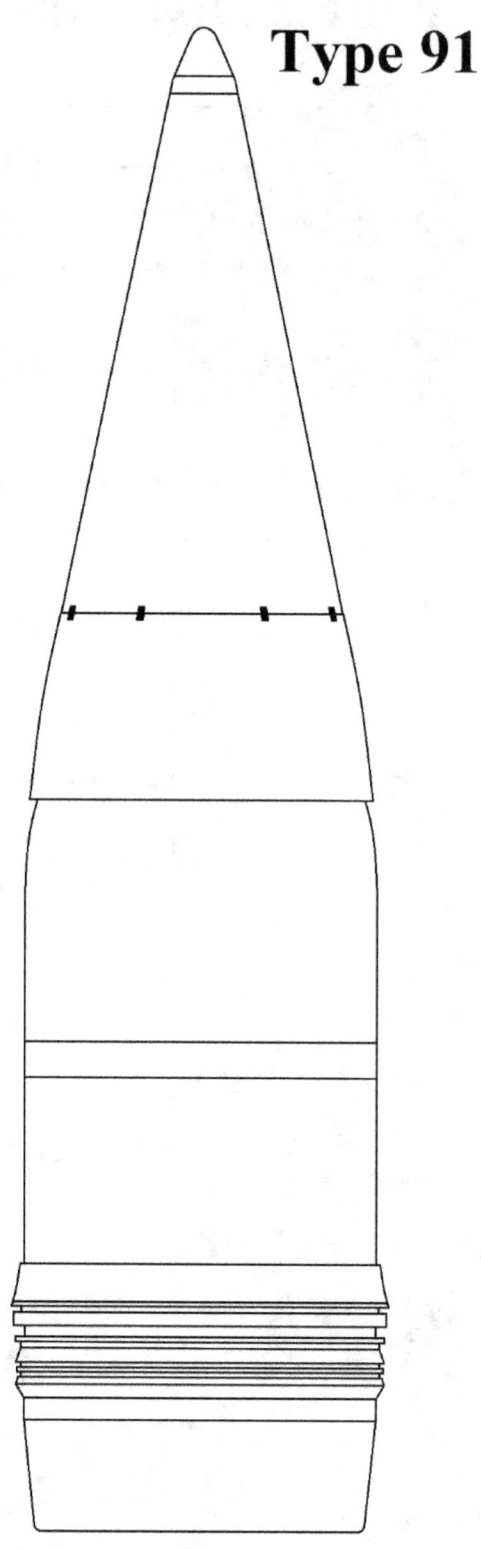 Type 91

Appendix M – Longitudinal Sections

Figure AM.1

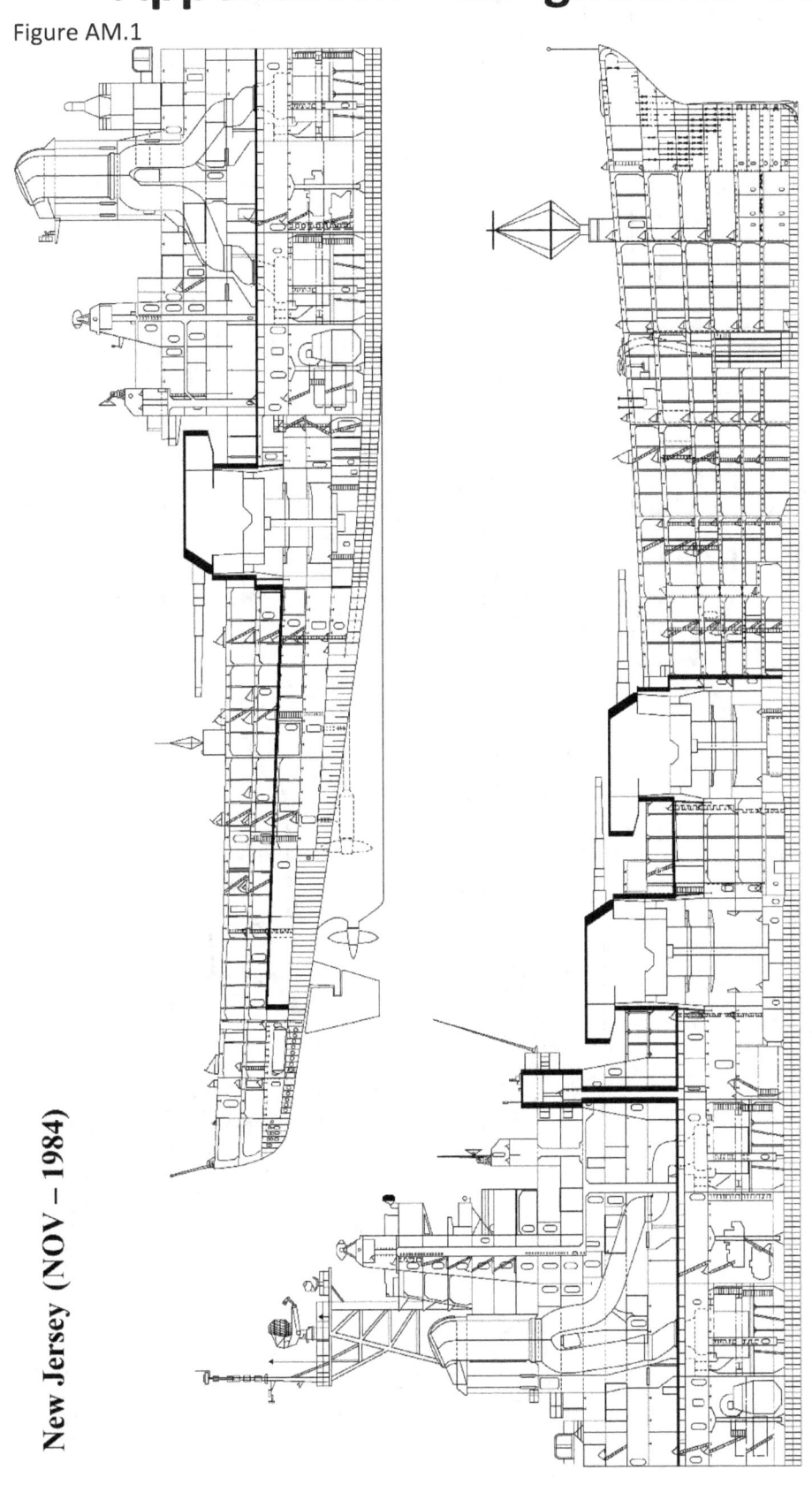

New Jersey (NOV – 1984)

Figure AM.2

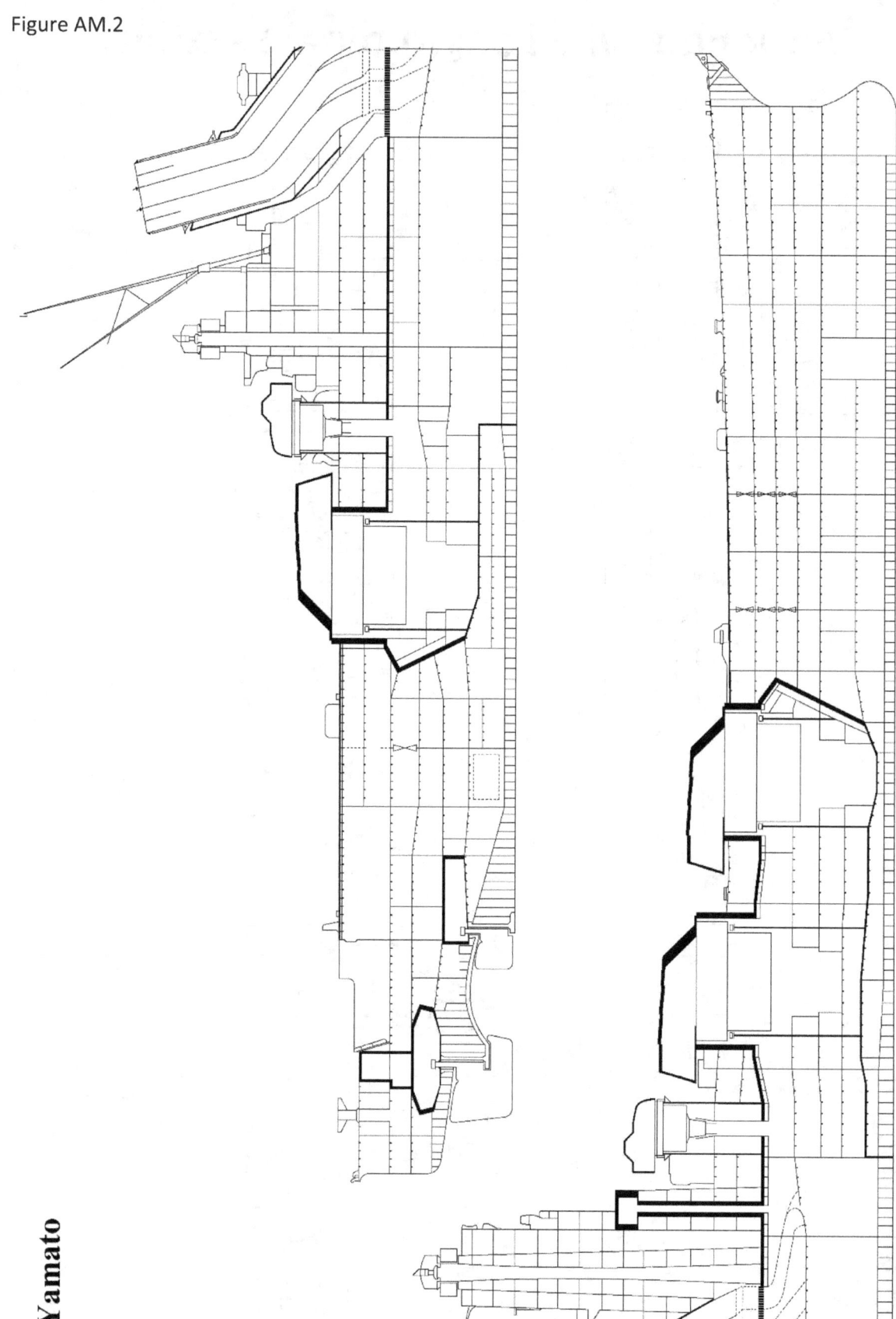

Yamato

Definitions

Angle of attack – The angle between the tangent to the trajectory of the projectile at the point of impact and the plate normal line through the point of impact.

Angle of descent – The vertical angle between the horizontal plane and the tangent to the trajectory of the projectile at the point of fall.

Angle of fall – Same as angle of descent.

Armoured citadel – The passive protective system formed by the main armour belt, the main armoured transverse bulkheads and the main armour deck.

Armour-piercing cap – Designed to enhance their armour penetration capacity and bite angle, armour-piercing projectiles are fitted with this kind of cap.

Armour-piercing projectile (AP) – Designed to defeat heavy armour, projectiles of this type have thick metallic walls and contain a comparatively limited amount of explosives.

Average gun – A used gun with a degree of bore enlargement corresponding to circa half the life of the barrel in terms of equivalent full charges (EFC) fired. Muzzle velocity, ballistic characteristics and vertical armour penetration capacity of an average gun compared to that of a new gun are inferior, but horizontal armour piercing capacity is increased due to more plunging trajectories.

Average pressure profile – Constant function having the same limit points and integral value as the function of the actual pressure inside the gun.

$\int p(x)dx = \int p_{av}(x)dx = p_{av} * L$,

whence

p=pressure,

p_{av}=average pressure,

L=Length of gun barrel.

Ballistic cap – Streamlined, hollow windshield provided to reduce retardation.

Ballistic characteristics – Characteristics dependent on initial velocity and the ballistic coefficient of the projectile, namely max. range, time of flight, max. ordinate, angle of fall, drift, danger space etc. The flatter the trajectory the better the ballistic characteristics of the gun. The higher the initial velocity and the more favorable the ballistic coefficient of the projectile, the flatter the trajectory.

Ballistic coefficient – Dependent on the sectional density and shape of the projectile, ballistic coefficient indicates the ability of the projectile to overcome the resistance of the medium it travels through. Retardation greatly depends on this coefficient.

Ballistic limit velocity (BL) – Terminal velocity required for a given projectile to perforate a plate of given type and thickness at a specified obliquity.

Barrel life – The life of a gun barrel is measured by the number of equivalent full charges (EFC) that can be fired without having to reline the gun.

Boat tail – Taper of the projectile behind the driving band as an attempt to enhance its ballistic coefficient.

Bore erosion – Amortization of the gun bore due to usage. The higher the average pressure profile of the gun, the more severe the erosion.

Bursting charge – Explosive filling of projectiles.

Caliber radius head (crh) – Measurement of how streamlined the head of a projectile is. Larger numbers indicate more pointed shell heads.

Capital ship – Armoured vessel of war, not an aircraft carrier, mounting a battery of a caliber greater than 8 inches (203 mm).

Cemented armour (C) – Armour plates of this type have hardened face in order to shatter the armour-piercing cap/body tip of AP projectiles. The backing layer of the hardened face has comparable metallurgical properties to those of homogeneous armour plates. Cemented armour is most efficient when impact angles are relatively close to the normal. Hence, heavy vertical armour was usually constructed of this material.

Class A – Cemented armour (US terminology)

Class B – Thick homogeneous armour (US terminology)

Danger space – The distance measured along the line of fire in front of the target that, if the target were moved toward the firing point, a shot striking the base of the target in its original position would strike the top of the target in its new position.

Decapped projectile – Capped armour piercing projectile whose armour piercing cap has been knocked off or shattered.

Density factor – Projectile weight divided by the cube of bourrelet diameter (m/d^3).

Detrimental hardening – Projectiles having this type of hardness distribution have virtually identical hardness on the surface and at the center. The British and the Japanese preferred this hardness distribution.

Director – Artillery position housing fire-control sensors.

Dispersion – The distance of the point of impact of the shot from the mean point of the impact of the salvo. Dispersion in range is measured along the line of fire and in deflection at right angles to the line of fire.

Drag – The effect of the resistance of the medium which the projectile travels through.

Dual-purpose gun (DP) – Gun capable of effectively engaging both surface and aerial targets.

Firing delay – Deliberate pause of a fraction of a second between the firing of individual shots to prevent interference.

Gun elevation – The vertical angle formed by the bore axis and the horizontal plane.

Hardness – Resistance of a material to localized deformation, indentation or penetration.

High explosive projectile (HE) – Designed to destroy lightly protected targets, projectiles of this type have relatively thin metallic walls and contain large quantities of explosives as compared to armour-piercing shells.

High-performance gun – Refers to guns with above-average muzzle energy for their respective bore diameter. High-performance guns have a high average pressure profile.

High-pressure gun – Refers to guns with an above-average average pressure profile. Guns having an average pressure profile of over 9.5 tons/in² are termed high-pressure guns in this study.

High-velocity gun – Guns with at least above-average muzzle velocity. Guns having an initial velocity of at least 2,600 fps (792 mps) are termed high-velocity guns in this study.

Homogeneous armour – Tough steel with theoretically uniform mechanical properties. This type of material is advantageous when high impact obliquities are anticipated.

Illuminating projectile – Used at night to illuminate the target, these projectiles contain an illuminant whose descent is haltered by a parachute.

Immunity zone – The period of impenetrability of the armoured citadel in terms of range (distance) against a specified attack. The inner limit of the immunity zone is secured by vertical armour. As range opens up, terminal energy of incoming projectiles decreases while their angle of fall increases. Consequently – if the vertical protection of the ship is strong enough to counter the specified threat – side armour becomes impenetrable beyond a certain range. The outer limit of the immunity zone is secured by horizontal armour. Albeit terminal energy decreases, impact angles against horizontal armour become more favorable for penetration as range opens up. Beyond a certain range – which marks the outer limit of the period of immunity – horizontal armour becomes vulnerable. The distance between the inner and outer limits is the zone where the ship is theoretically secured against the specified threat. The immunity zone constantly oscillates as the ship is pitching, rolling, its speed and relative bearing changes etc., which is why a too small period of immunity – in terms of distance – is inadequate to provide reliable protection. Ideally, armoured ships seek to engage at ranges corresponding to the center of their immunity zone. When the immunity zone of an armoured ship was calculated, the specified threat was often the gun carried by the ship itself, as the performance of foreign guns was not known in detail.

Impact angle – Same as angle of attack.

Initial velocity (IV) – Velocity of the projectile at the point of leaving the muzzle.

Kinetic energy upon impact – Kinetic energy of the projectile at the point of impact.

Low-pressure gun – Refers to guns with below-average average pressure profile. Guns having an average pressure profile of less than about 9.0 tons/in^2 are termed low-pressure guns in this study.

Low-velocity gun – Refers to guns with below-average muzzle velocity. Guns having an initial velocity of no more than about 2,500 fps (762 mps) are termed low-velocity guns in this study.

Main battery – Guns of the largest caliber carried by the ship.

Maximum ordinate – Highest point of the trajectory.

Muzzle energy – Kinetic energy of the projectile at the point of leaving the muzzle. $E_k = \int p(x)dx \cdot d^2 \cdot \pi/4 = p_{av} \cdot L \cdot d^2 \cdot \pi/4 = m \cdot v^2/2$,

whence

E_k = kinetic energy,
d = diameter of gun barrel,
m = mass of projectile,
v = velocity of projectile,
p = pressure,
p_{av} = average pressure,
L = Length of gun barrel.

Muzzle velocity – Same as initial velocity.

New gun – An unused gun with no bore enlargement propelling shells at designed muzzle velocity.

Non-cemented armour (NC) – Same as homogeneous armour.

Obliquity – Same as angle of attack.

Ogive – Forward curved section of a projectile limited by its shoulder.

Period of immunity – Same as immunity zone.

Piercing – Failure mode of elastoplastic materials (homogeneous armour) against sharp rigid penetrators.

Plugging – Failure mode of elastoplastic materials (homogeneous armour) against blunt rigid penetrators.

Pressure profile – Pressure as a function of distance or time inside the gun barrel.

Range – The distance from a station on one's own ship to the target or some other designated point.

Rate of fire – Rounds fired as a function of time.

Relative effectiveness factor (RE) – Dimensionless explosive constant indicating the potency of explosives as compared to TNT, which has a RE factor of 1.00. Higher RE indicates greater destructive potential. The relation between RE and potency is directly proportional, i.e. RE factor indicates the relative mass of TNT to which an explosive is equivalent.

Relative target bearing – The bearing of the target from the firing ship measured in the horizontal from the bow of one's own ship clockwise from 0 deg. to 360 deg.

Retardation – Velocity loss during the flight of the projectile due to the drag force of the medium it travels through.

Ricocheting – The tendency of projectiles to bounce off the target beyond a certain angle of impact, which varies with each individual attack/plate combination.

Salvo – Two or more shots fired simultaneously.

Secondary armament – In the case of capital ships, all guns except for those of the largest caliber.

Sectional density – Projectile weight divided by the square of bourrelet diameter (m/d^2).

Semi-armour piercing projectile (SAP) – Designed to defeat moderately armoured targets, projectiles of this type have thicker metallic walls and carry less explosives as compared to HE type projectiles, but have thinner walls and carry more explosives as compared to AP type projectiles.

Sheath hardening – Projectiles having this type of hardness distribution have greater hardness on the surface than at the center. The surface hardness of sheath hardened projectiles drops markedly only abreast their cavity. The Americans and the Germans preferred this hardness distribution.

Shore hardness test – Test measuring the hardness of materials.

Striking velocity – Velocity of the projectile at the point of impact.

Target angle – The relative bearing of one's own ship from the target, measured in the horizontal plane from bow of the target clockwise from 0 deg. to 360 deg.

Terminal energy – Same as kinetic energy upon impact.

Terminal velocity – Same as striking velocity.

Time of flight – Elapsed time between the projectile leaving the muzzle and hitting the target.

Trajectory – The path that an object with mass in motion follows through space as a function of time.

Vertex – Same as maximum ordinate.

List of Tables

Table No.	Description	Page
Table 1.1	Armament Comparison	15
Table 1.2	Broadside Comparison	16
Table 1.3	Comparison of Ballistic Characteristics	16
Table 1.4	Time of Flight Comparison	16
Table 1.5	Comparison of Armour Penetration Capacity Based on Bu. Ord. Sk. 78841	17
Table 1.6	Comparison of Armour Penetration Capacity Based on Official Data	17
Table 1.7	Secondary Armament Comparison	26
Table 2.1	Limits of Type 98 System	29
Table 2.2	Fire-Control Equipment Comparison	31
Table 3.1	Protection – Iowa	49
Table 3.2	Protection – Yamato	50-51
Table 3.3	Dimensions – Iowa	51
Table 3.4	Dimensions – Yamato	51
Table 3.5	Iowa's Immunity Vs. Yamato – I.	52
Table 3.6	Iowa's Immunity Vs. Yamato – II.	52
Table 3.7	Yamato's Immunity Vs. Iowa	52
Table C.1	Best-Case Vs. Worst-Case Scenario	77
Table AA.1	General Characteristics	83
Table AB.1	Decapping	86
Table AC.1	Japanese Armour (ATC)	91
Table AC.2	Japanese Armour (NPG)	92
Table AD.1	U.S. Armour (NPG+ATC)	97
Table AE.1	Flat Penetrators – Summary of Test Results (NPS 7-43)	100
Table AE.2	Flat Penetrators – Summary of Test Results (China Lake, California - December 1973)	100
Table AG.1	Maximum Range Factors (Radars)	103
Table AG.2	Visual and Radar Horizon	103
Table AG.3	Fire-Control Radars	104
Table AG.4	Surface Search Radars	104
Table AG.5	Air Search Radars	105
Table AG.6	Iowa's Air Search Radar	115
Table AG.7	Iowa's Surface Search Radar	116
Table AG.8	Yamato's Air Search Radars	116

List of Figures

Figure No.	Description	Page
Figure 1.1	Main Gun Comparison	18
Figure 1.2	Broadside Comparison	19
Figure 1.3	Kinetic Energy Comparison	20
Figure 1.4	Iowa's Armour Penetration Capacity	21
Figure 1.5	Yamato's Armour Penetration Capacity	22
Figure 1.6	Armour Penetration Capacity Comparison – I.	24
Figure 1.7	Armour Penetration Capacity Comparison – II.	25
Figure 2.1	Iowa's Main Battery Fire–Control System	32-35
Figure 2.2	Yamato's Main Battery Fire–Control System	36-40
Figure 2.3	Dispersion and Danger Space of Guns, Range Accuracy of Radars, Rangefinder's Unit of Error	41
Figure 2.4	Visual & Radar Horizon	42
Figure 3.1	Longitudinal Sections	53-54
Figure 3.2	Cross Sections	55-63
Figure 3.3	Iowa's Immunity Vs. Yamato – I.	64-65
Figure 3.4	Yamato's Immunity Vs. Iowa – I.	66-67
Figure 3.5	Immunity Zones at 90 deg.	68
Figure 3.6	Iowa's Immunity Vs. Yamato – II.	69
Figure 3.7	Yamato's Immunity Vs. Iowa – II.	70
Figure 3.8	Immunity Comparison – I.	71
Figure 3.9	Immunity Comparison – II	72
Figure C.1	Immunity Graphs – Assumed Vs. Actual	79
Figure C.2	Immunity Graphs – Bu. Ord. Sk. 78841 Vs. DeMarre – I.	80
Figure C.3	Immunity Graphs – Bu. Ord. Sk. 78841 Vs. DeMarre – II.	81
Figure AB.1	Iowa's Armour Penetration Curves	88-89
Figure AC.1	Iowa's Guns Vs. 26" Turret Face	94
Figure AC.2	Hardness Distribution of 26" Turret Face	95
Figure AF.1	Soft Vs. Hard Penetrator	102
Figure AG.1	Radar+IFF System	106
Figure AG.2	U.S. Radars	107-114
Figure AH.1	Mark 37 GFCS	117-120
Figure AI.1	Yamato's Armour Decks	121-122
Figure AJ.1	U.S. Battleship's Combat Information Center (CIC)	123
Figure AK.1	Yamato's Main Turrets	125-127
Figure AL.1	Iowa's AP Projectile	129-131
Figure AL.2	Yamato's AP Projectile	132-134
Figure AM.1	Iowa's Longitudinal Section	135
Figure AM.2	Yamato's Longitudinal Section	136

List of Pictures

Picture No.	Description	Page
Cover	Yamato running machinery trials off Bungo Strait on 20 October 1941	---
Picture 1.1	Exquisite view of USS New Jersey, 11 September 1968.	12
Picture 1.2	The gargantuan forward turrets of Musashi, photographed in 1942.	14
Picture 2.1	Iowa's aft Mark 37 (5"/127 mm) and Mark 38 (16"/406 mm) directors, captured on 8 October 1983. The World War II period Mark 4 and Mark 8 fire-control radars had been replaced by the more advanced Mark 25 and Mark 13 sets, respectively. Notice the protruding hoods of the rangefinders.	28
Picture 2.2	Excellent view of the fire-control tower of Musashi, captured in 1942. The rangefinder hoods of the main and secondary turrets, as well as that of the foretop director, can be clearly seen.	30
Picture 3.1	17.3" (439 mm) side armour of New Jersey's conning tower. The picture was taken on 1 July 1981.	45
Picture 3.2	Yamato running full-power trials in Sukumo Bay, 30 October 1941. Notice the upper edge of the main belt of the ship.	47
Picture AC.1	26" (660 mm) armour plate, previously attacked by 2,700 lbs/16" (1,225 kg/406 mm) AP projectile at 90 deg. Limit velocity was established at 1,839 fps (561 mps) +/-3 %. Selfsame armour plates protected the frontal side of Yamato's main turrets, but they were sloped 45 deg. from the vertical. NPG 5-47 describes the actual turret face plates as impenetrable at any range due to the combination of extraordinary thickness and inclination. This conclusion is in agreement with our calculations.	93

Bibliography

Books
Dulin Jr., Robert O. and Garzke Jr., William H., et al. *Battleships: United States Battleships in World War II.* Naval Institute Press; 1st edition (January 1, 1976) 978-0870210990

Dulin Jr., Robert O. and Garzke Jr., William H., et al. *Battleships: Axis and Neutral Battleships in World War II.* Naval Institute Press; 1st edition (November 27, 1985) 978-0870211010

Campbell, N.J.M. *Naval Weapons of World War II.* Naval Inst. Pr.; 1st edition (January 1, 1985) 978-0870214592

Sumrall, Robert F. and Walkowiak, Thomas. *Iowa Class Battleships: Their Design, Weapons and Equipment.* Naval Inst Pr; Reprint edition (April 1, 1989) 978-0870212987

Skulski, Janusz. *Battleships Yamato and Musashi (Anatomy of The Ship).* Osprey Publishing (January 23, 2018) 978-1472832245

Lengerer, Hans and Ahlberg, Lars. *Capital Ships of the Imperial Japanese Navy 1868-1945: The Yamato Class and Subsequent Planning.* Nimble Books (December 16, 2014) 978-1608880836

Bureau of Naval Personnel. *Naval Ordnance and Gunnery.* Periscope Film LLC; Illustrated edition (May 25, 2013) 978-1937684228

Naval Education & Training Center. *Navy Electricity & Electronics Training Series: Module 18 - Radar Principles - Navedtra 14190 - (Nonresident Training Course).* lulu.com; null edition (July 13, 2013) 978-1304228741

Articles
Morss, Strafford. *Iowa vs. Yamato.* Warship International Vol. 23, No. 2 (1986), pp. 118-136, International Naval Research Organization

Morss, Strafford. *The Washington Naval Treaty and the Armor and Protective Plating of USS "Massachusetts": Part I.* Warship International Vol. 43, No. 3 (2006), pp. 273-309, International Naval Research Organization

Morss, Strafford. *The Washington Naval Treaty and the Armor and Protective Plating of USS "Massachusetts", Part II.* Warship International Vol. 43, No. 4 (2006), pp. 395-435, International Naval Research Organization

Admiralty Minutes
ADM 213/378 – ATC Meeting 15th August 1946

ADM 281/37 – Admiralty Armour Investigation Program (1946-1950), Investigation No. 10, The Ballistic Performance of Cemented and Face Hardened Deck Armour (200 lbs/sq. ft. and 240 lbs/sq. ft.) Under Attack by Decapped A.P.C. Shell

ADM 281/126 – ATC Meeting 20th November 1947

ADM 281/127 – ATC Meeting 22nd July 1948

Unites States Naval Technical Mission to Japan
USNTMJ OT O-19 – Japanese Projectiles General Type
USNTMJ OT O-45 (N) – Japanese 18" Guns and Mounts
USNTMJ OT O-31 – Japanese Surface and General Fire Control
USNTMJ OT O-29 – Japanese Fire Control

Naval Proving Ground (Dahlgren, Virginia)
NPG 2-43 – The Effects of Nose Shape on the Ballistic Performance of 15 lb. 3" AP Solid Shot Against Homogenous Armor Plate
NPG 7-43 – Penetration of Homogenous Armor by 3-inch Flat-Nosed Projectile
NPG 4-44 – The Penetration of Homogenous Plate at Various Obliquities
NPG 3-47 – Metallurgical Examination of Standard U.S. Armor-Piercing Projectiles
NPG 5-47 – Ballistic Tests and Metallurgical Examination of Japanese Heavy Armor Plate

Ordnance Pamphlets
O.P. 770 – 16-Inch Range Table
O.P. 658 – Fire-Control Radar, Mark 8

Bulletins
Radar Operator's Manual, Radar Bulletin No. 3 (April 1945)

Booklet of General Plans
BB62 New Jersey (NOV – 1984, Long Beach Naval Shipyard, Long Beach, California)

Other Documents
Ipson, Thomas W. *Effect of Projectile Nose Shape on Ballistic Limit Velocity, Residual Velocity, and Ricochet Obliquity.* (January 1, 1973)
U.S. Navy Projectiles and Fuzes (June 1945)

Online
http://www.navweaps.com

Other Books in this Series

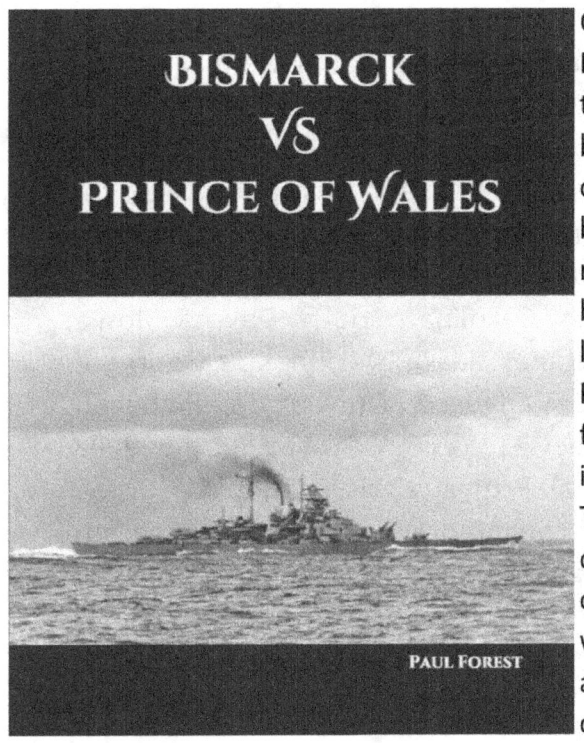

On 24 May 1941 the German battleship Bismarck engaged in a gunnery duel with the flagship of the Royal Navy, the battlecruiser HMS Hood, and the newly commissioned King George V class battleship, HMS Prince of Wales. Within minutes, Hood found a watery grave beneath the devastating shellfire of the legendary German ship. Not much later, Prince of Wales made smoke and left the field of battle. But would it have been indeed so foolhardy to fight Bismarck alone? The purpose of this concise inquiry is to compare the relevant technical characteristics of the two ships and discover whether Prince of Wales alone could have achieved victory and, if so, under what conditions.

Littorio and Richelieu were commissioned during the spring of 1940. Although the imminent Fall of France prevented the envisaged struggle for the control of the Mediterranean between this nation and Italy, the climax of this campaign would have been undoubtedly a gunnery duel between the newly commissioned battleships. We now endeavor to investigate which ship would have been more likely to emerge victorious from this hypothetical engagement.

Available on Amazon.com in Kindle eBook and paperback formats.

Also by this Author

This comprehensive reference book contains detailed tabulations and line drawings of the protection schemes and immunity graphs of Allied battleships and battlecruisers designed after the Washington Naval Treaty, including the following capital ship classes:

North Carolina, South Dakota, Iowa, Montana, Alaska, Nelson, King George V, Lion, Vanguard, Dunkerque, Richelieu, Project 21, Type A, Project 23 (Sovetsky Soyuz), Project 23bis, Project 24, Project 69 (Kronshtadt), Project 82 (Stalingrad), Design 1047;

and fragmentary information on UP 41, Project 23UN, Project 25, Project 64.

This comprehensive reference book contains detailed tabulations and line drawings of the protection schemes and immunity graphs of Axis battleships and battlecruisers designed after the Washington Naval Treaty, including the following capital ship classes:

Littorio, Scharnhorst, Bismarck, H-39, O, Yamato;

and fragmentary information on, H-40, H-41, H-42, H-43, H-44 and B-65.

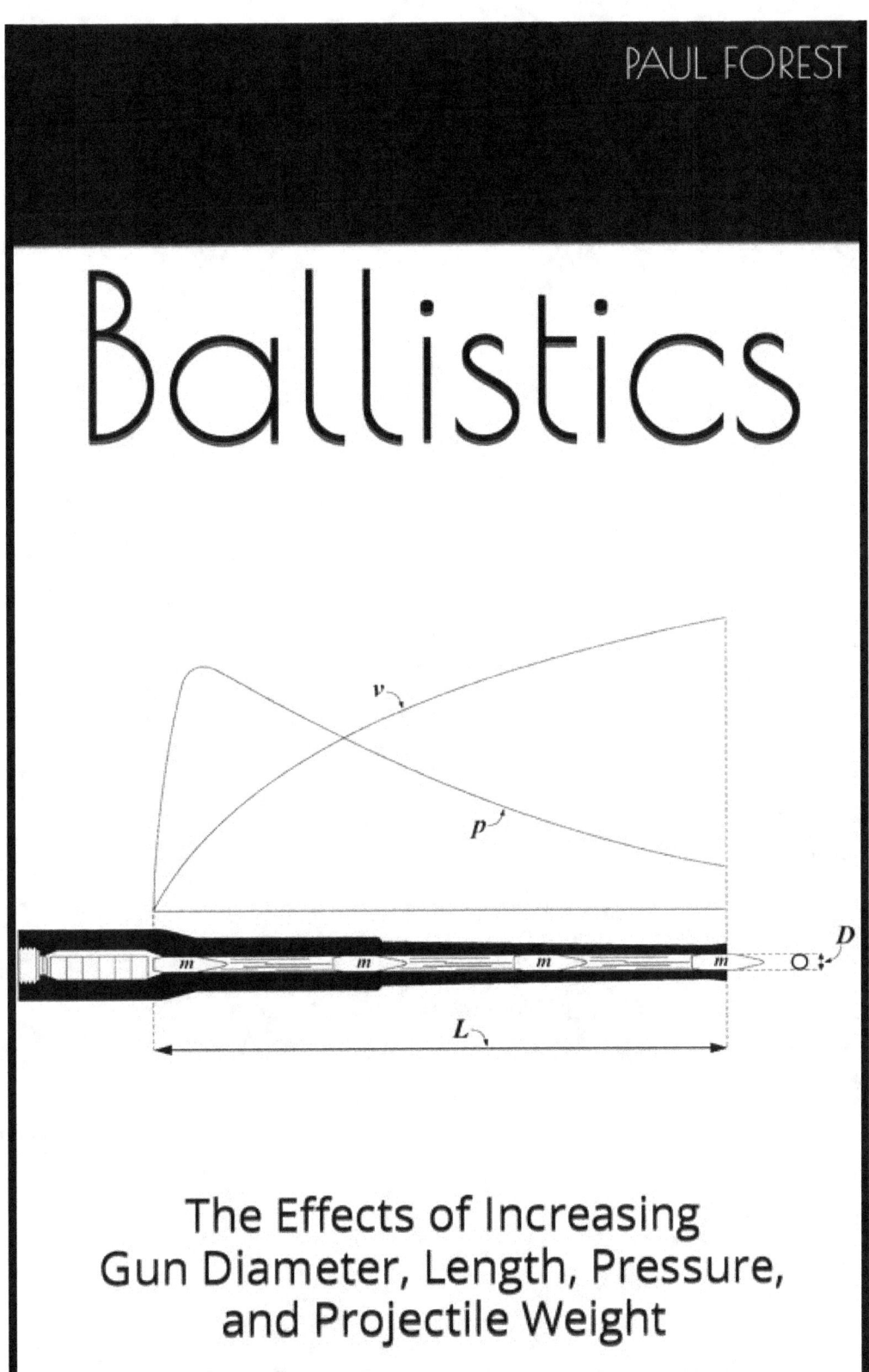

Ballistics

The Effects of Increasing Gun Diameter, Length, Pressure, and Projectile Weight

This short essay investigates the effects of increasing the diameter, length and pressure of large caliber naval guns, and the weight of projectiles, on ballistic characteristics and armour penetration capacity.

www.ingramcontent.com/pod-product-compliance
Lightning Source LLC
Chambersburg PA
CBHW060416220526
45465CB00008B/2902